Essentials of LTE and LTE-A

This practical, one-stop guide will quickly bring you up-to-speed on LTE and LTE-Advanced (LTE-A). With everything you need to know about the theory and technology behind the standards, this is a must-have for engineers and managers in the wireless industry.

- First book of its kind describing technologies and system performance of LTE-A
- Covers the evolution of digital wireless technology, basics of LTE and LTE-A, design of downlink and uplink channels, multi-antenna techniques, and heterogeneous networks
- Analyzes performance benefits over competing technologies, including WiMAX and 802.16m
- Reflects the latest LTE Release-10 standards
- Includes numerous examples, including extensive system and link results
- Unique approach is accessible to technical and non-technical readers alike

AMITABHA GHOSH is a Senior Director and Fellow of the Technical Staff at Motorola Solutions, where he works in the area of current and future air-interface technologies for 802.16m, 3GPP LTE, LTE-Advanced, and other broadband technologies. Since joining Motorola, he has worked on eight different wireless technologies, and is currently leading Motorola's efforts in defining 3GPP LTE and LTE-Advanced physical layer standards from the concept phase to the adopted baseline.

RAPEEPAT RATASUK is currently a Distinguished Member of the Technical Staff at Motorola Solutions. He has extensive experience in 3G/4G cellular system design and analysis (specifically LTE, HSPA, WiMAX, 1xEV-DV, and W-CDMA technologies), including algorithm development, performance analysis and validation, physical layer modeling, and simulations.

The Cambridge Wireless Essentials Series

Series Editors
WILLIAM WEBB *Neul, Cambridge, UK*
SUDHIR DIXIT *Nokia, US*

A series of concise, practical guides for wireless industry professionals.

Martin Cave, Chris Doyle and William Webb, *Essentials of Modern Spectrum Management*
Christopher Haslett, *Essentials of Radio Wave Propagation*
Stephen Wood and Roberto Aiello, *Essentials of UWB*
Christopher Cox, *Essentials of UMTS*
Steve Methley, *Essentials of Wireless Mesh Networking*
Linda Doyle, *Essentials of Cognitive Radio*
Nick Hunn, *Essentials of Short-Range Wireless*
Amitabha Ghosh and Rapeepat Ratasuk, *Essentials of LTE and LTE-A*

Forthcoming
Abhi Naha and Peter Whale, *Essentials of Mobile Handset Design*
Barry G. Evans, *Essentials of Satellite Communications*
David Bartlett, *Essentials of Positioning and Location Technology*

For further information on any of these titles, the series itself and ordering information see www.cambridge.org/wirelessessentials

Essentials of LTE and LTE-A

Amitabha Ghosh and Rapeepat Ratasuk
Motorola Solutions

CAMBRIDGE
UNIVERSITY PRESS

CAMBRIDGE UNIVERSITY PRESS
Cambridge, New York, Melbourne, Madrid, Cape Town,
Singapore, São Paulo, Delhi, Tokyo, Mexico City

Cambridge University Press
The Edinburgh Building, Cambridge CB2 8RU, UK

Published in the United States of America by Cambridge University Press, New York

www.cambridge.org
Information on this title: www.cambridge.org/9780521768702

© Cambridge University Press 2011

First published 2011

Printed in the United Kingdom at the University Press, Cambridge

A catalogue record for this publication is available from the British Library

Library of Congress Cataloguing in Publication data
Ghosh, Amitabha, 1958–
Essentials of LTE and LTE-A / Amitabha Ghosh and Rapeepat Ratasuk.
    p.   cm. – (The Cambridge wireless essentials series)
Includes bibliographical references.
ISBN 978-0-521-76870-2
1. Long-Term Evolution (Telecommunications)   I. Ratasuk, Rapeepat.   II. Title.
TK5103.48325.G485   2011
621.382–dc22

                                                                    2011011253

ISBN 978-0-521-76870-2 Hardback

To my parents for their continuous support and teaching me the value of education and thirst for knowledge; and to my family, Chittarupa, Devika, and Adit, for their support, encouragement, and love.

Amitabha Ghosh

To Tanita, Alisa, and Paul.

Rapeepat Ratasuk

Contents

Preface

The next-generation wireless broadband technology is changing the way we work, live, learn, and communicate through effective use of state-of-the-art mobile broadband technology. The packet-data-based revolution started around 2000 with the introduction of 1x Evolved Data Only (1xEV-DO) and 1x Evolved Data Voice (1xEV-DV) in 3GPP2 and High Speed Downlink Packet Access (HSDPA) in 3GPP. The wireless broadband fourth-generation technology (4G) is an evolution of the packet-based 3G system and provides a comprehensive evolution of the Universal Mobile Telecommunications System specifications so as to remain competitive with other broadband systems such as 802.16e (WiMAX). Specification work was started in late 2004 on Long Term Evolution (LTE) of the UMTS Terrestrial Radio Access and Radio Access Network intended for commercial deployment in 2010. Two main components constitute the LTE system architecture – the Evolved Universal Terrestrial Radio Access Network (E-UTRAN) and the Evolved Packet Core (EPC). The goals for the evolved system (E-UTRAN and EPC) included support for improved system capacity and coverage, high peak data rates, low latency, reduced operating costs, multi-antenna support, flexible bandwidth operations, and seamless integration with existing systems. The standardization work for LTE Rel-8 was completed in early 2009 and commercial LTE systems will be deployed in the 2011–2012 timeframe. LTE Rel-8 is currently evolving to LTE-Advanced (LTE Rel-9 and Rel-10), which will further improve the spectral efficiency, peak rates, and user experience compared with LTE Rel-8. LTE-Advanced has also been approved by the International Telecommunication Union (ITU) as an International Mobile Telecommunications-Advanced (IMT-A) technology.

The book is organized in seven chapters. Chapter 1 gives a timeline and brief description of the evolution of digital wireless technology starting with GSM, IS-95, cdma2000 1x, WCDMA Rel-99, HSPA (Rel-5/6),

WiMAX, LTE, LTE-Advanced, and 802.16m with emphasis on how supported data rates, throughput, and applications have evolved. Chapter 2 provides a brief description of LTE requirements and system architecture together with the basic principles of orthogonal frequency-division multiple-access (OFDMA) and single-carrier frequency-division multiple-access (SC-FDMA) technology. Chapter 3 dives into the basic details of LTE downlink OFDMA transmission including transport and physical-channel structure, control-channel details, system operations, and inter-cell interference coordination schemes both for FDD (Frequency-Division Duplex) and for TDD (Time-Division Duplex) LTE. Aspects of downlink system performance under various channels and antenna structure are summarized at the end of the chapter.

Chapter 4 provides the details of LTE uplink transport and physical-channel structure, control-channel details, random access, system operations, and fractional power control followed by uplink system performance under various channels and antenna configurations. The LTE system offers a rich suite of multiple-antenna techniques that can be used in various scenarios to improve the performance and user experience. Chapter 5 describes various multi-antenna schemes for LTE downlink and uplink and provides a system-performance comparison of various multi-antenna schemes. Chapter 6 is devoted to technologies for LTE-Advanced (LTE-A). The chapter describes the requirements for IMT-A and how LTE-A will satisfy those requirements using enhanced technologies. The technologies include support of wider bandwidth using carrier aggregation, uplink spatial multiplexing, enhanced downlink spatial multiplexing, coordinated multiple-point transmission and reception, and heterogeneous networks including relays, distributed antenna systems, and pico-cells. Aspects of the system performance of these enhancements are presented and compared with the performance of the legacy LTE system.

Finally, Chapter 7 provides a comparison of LTE/LTE-A with other competitive broadband systems such as 802.16e/802.16m. As the name signifies, this chapter outlines both qualitative and quantitative differences between the 802.16e/802.16m (WiMAX) system and the LTE/LTE-A system. System performance comparisons between these systems are presented for various reuse schemes and antenna configurations.

At the time of writing, there are ongoing discussions within the operator and vendor community regarding further evolution of LTE-A technology. These enhancements will appear in Rel-11 and Rel-12 of 3GPP and will offer better user experience, lower cost per bit, greener base stations, and efficient self-organizing networks.

Acknowledgments

Several of our colleagues made a significant impact on the materials presented in this book. We would like to acknowledge and thank Prakash Moorut for his comments and suggestions on the spectrum-engineering aspects, Bishwarup Mondal, who provided critical comments, simulations, and suggestions for improving the contents related to multi-antenna systems and heterogeneous networks, Nitin Mangalvedhe for providing us with some of the simulation results and his in-depth comments relating to heterogeneous networks, Joe Hoffman for providing help related to the economic aspect of wireless broadband systems, Mark Cudak for providing us with his expertise on WiMAX-related issues, and Tim Thomas, who reviewed the entire first draft of the book and provided constructive comments and criticisms. Throughout our professional careers at Motorola we had the good fortune of working and learning from some of the most talented people in the cellular industry, including Ken Stewart, Bob Love, the late Dennis Schaeffer, Fan Wang, Joe Pedziwiatr, Paul Steinberg, Phil Fleming, Fred Vook, Weimin Xiao, Brian Classon, and 3GPP colleagues, among many others. Finally, we would like to thank our superiors Sudhakar Ramakrishna and Bill Payne for providing us with encouragement and support for undertaking this project.

1　Genesis of wireless broadband technology (from 2G to 4.5G)

1.1 Genesis of wireless technology

The digital cellular technology revolution started with the introduction of GSM (Groupe Special Mobile) in the late 1980s. The GSM technology was based on time-division multiple access (TDMA) and was capable of supporting data services of up to 9.6 kbps. In the early 1990s, IS-95, a standard based on code-division multiple-access (CDMA) technology was introduced. This offered data rates of up to 14.4 kbps and improved spectral efficiencies over a GSM system. Subsequently, both these technologies evolved over time, with each phase offering higher peak rates and improved sector/edge spectral efficiencies. Both GSM and IS-95 CDMA evolved in different phases. In 1997, the Generalized Packet Radio System (GPRS) based on packet data instead of circuit data was standardized, followed by Enhanced Data Rates for Global Evolution (EDGE). Also, at the end of 1998, the Third-Generation Partnership Project (3GPP) was started. This was responsible for defining a third-generation (3G) wideband CDMA (WCDMA) standard based on the evolved GSM core network. At the same time the GSM standardization work was moved from ETSI SMG2 to 3GPP, and was called GERAN. Similarly, in the United States the IS-95 standard evolved to cdma2000 under the umbrella of Third-Generation Partnership Project 2 (3GPP2).

The packet-data-based revolution started around 2000 with the introduction of cdma2000 1x Evolved Data Only (1xEV-DO) and 1x Evolved Data Voice (1xEV-DV) in 3GPP2 and High Speed Downlink Packet Access (HSDPA) in 3GPP. These 3.5G technologies had the following common attributes: adaptive modulation and coding, hybrid automatic repeat request, fast scheduling based on smaller frame size, turbo codes, and de-centralized architecture to reduce latency. In the next phase of development of 3.5G technology, improved uplink functionality was added to 3GPP and 1xEV-DO systems. Concurrently, advances were made in cdma2000 1x technology (i.e. cdma 1x-advanced), which included an advanced vocoder, mobile receive diversity, an advanced receiver with interference cancellation, and advanced power control. It may be noted that, although 1xEV-DV was standardized, it never took off as a technology due to the reluctance of the operator community to adopt the technology and the absence of proper eco-systems.

A disruptive technology known as mobile WiMAX based on orthogonal frequency-division multiplexing (OFDM) technology was standardized in 2006, and was dubbed the first 4G multiple access system. This technology was based on the IEEE 802.16e standard and offered scalable bandwidth up to 20 MHz, higher peak rates, and better spectral efficiencies than those provided by 3.5G systems. With the emergence of packet-based wireless broadband systems such as WiMAX, it was evident that a comprehensive evolution of UMTS would be required in order for it to remain competitive in the long term. As a result, work began on Evolved UMTS Terrestrial Radio Access (E-UTRA) based on the OFDM air interface. The Long Term Evolution (LTE Rel-8) system supports high peak data rates and provides low latency, improved system capacity and coverage, reduced operating costs, efficient multi-antenna support, efficient support for packet data transmission, flexible bandwidth of up to 20 MHz, and seamless integration with existing systems. The CDMA-based HSPA technology is also being enhanced to support quad carriers (bandwidth up to 20 MHz), MIMO, and higher-order modulation both on the downlink and on the uplink. A 4G proposal called Ultra Mobile Broadband (UMB) based on OFDM was also adopted by 3GPP2, but it failed to make any impact.

Both WiMAX and LTE are currently being enhanced (LTE-Advanced and 802.16m) so as to support even higher peak rates, higher throughput and coverage, and lower latencies resulting in a better user experience. Further, LTE-Advanced and 802.16m also enable one to meet or exceed IMT-Advanced requirements. Finally, the 4.5G wireless broadband systems will be standardized in 3GPP Rel-12 in the 2013–2017 timeframe. It is clear that 4.5G systems will further enhance the 4G systems in terms of user experience, sector spectral efficiency, and peak rates, but the exact features for 4.5G systems are still being decided.

The Digital Video Broadcasting (DVB) standards, which include Mediaflow and Multimedia Broadcast Multicast Service (MBMS) designed for LTE and HSPA, for global delivery of broadcast services such as digital television are also evolving to provide better spectral efficiencies for broadcast services.

The wireless evolution chart of 2G to 4.5G technology migration is shown in Figure 1.1.

The downlink peak rate improvement on going from 2G to 4.5G technology is shown in Table 1.1.

The improvement in downlink sector spectral efficiencies on going from 2G to 4.5G systems is shown in Figure 1.2.

It may be observed from Figure 1.2 that there has been an improvement by a factor of 30 in sector spectral efficiency with 4G systems compared with 2G, which results in improved cost per bit. Figure 1.3 shows an example of how mobile broadband cost per bit decreases exponentially with technology innovation in wireless technology.

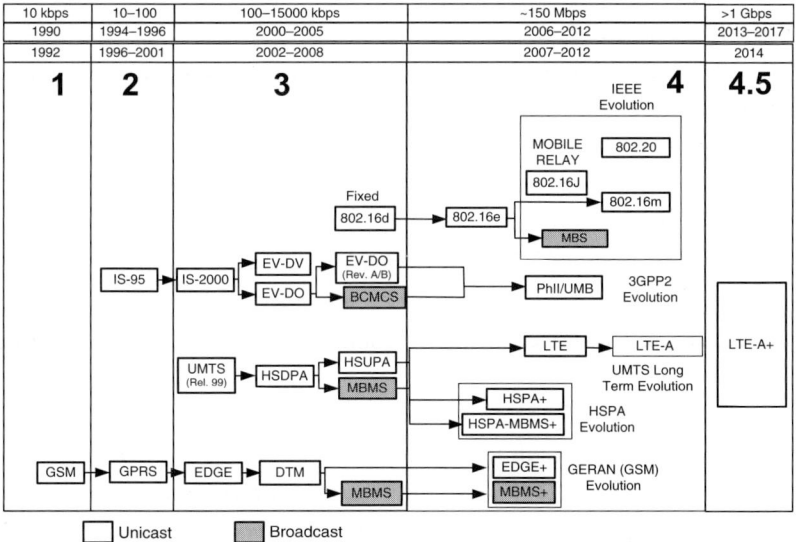

Figure 1.1. Standards evolution of wireless technologies (from 2G to 4.5G).

Table 1.1. *Downlink peak rates for different technologies*

Technology	Theoretical peak rates
GSM (2G)	9.6 kbps
IS-95 (2G)	14.4 kbps
GPRS (2G)	171.2 kbps
EDGE (2.5G)	473 kbps
cdma2000 1x (2G)	628.4 kbps
WCDMA (3G)	1920 kbps
GERAN/EGPRS2 (3G)	947.2 kbps
HSDPA Rel-5 (3.5G)	14 Mbps
cdma2000 1xEV-DO (3G)	3.1 Mbps
HSPA Rel-9 (3.5G)	84 Mbps (2×2 MIMO, Dual Carrier)
LTE Rel-8 (4G)	300 Mbps (20 MHz, 4×4 MIMO)
WiMAX (4G)	26 Mbps (10 MHz, 2×2 MIMO)
WiMAX/802.16m (4.5G)	303 Mbps (20 MHz, 8×8 MIMO)
LTE-Advanced Rel-10 (4.5G)	3 Gbps(100 MHz, 8×8 MIMO)

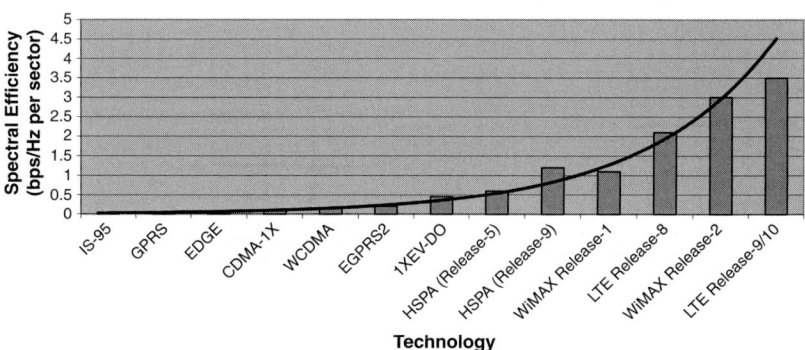

Figure 1.2. Improvement in downlink spectral efficiency going from 2G to 4G systems.

1.2 Key drivers for 4G/4.5G wireless broadband

Technology cycles tend to last on average 10 years. Thus, we have seen mainframe computing (1960s), minicomputing (1970s), personal computing (1980s), desktop internet computing (1990s), and finally mobile internet computing in the 2000s [1]. The need for 4G systems such as LTE

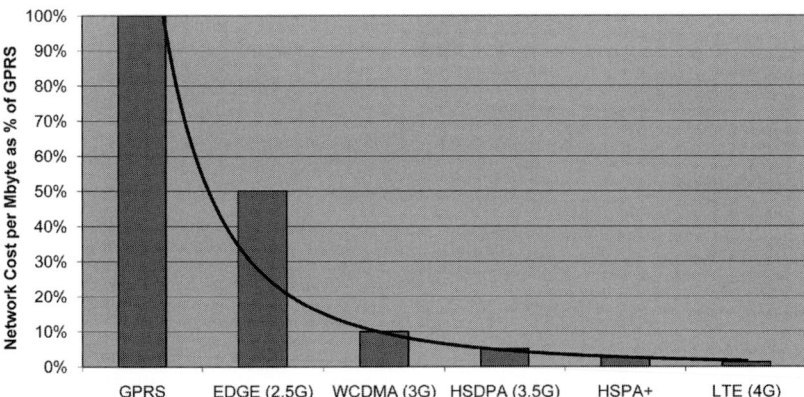

Figure 1.3. Mobile data cost per bit as a function of technology (adapted from [1]).

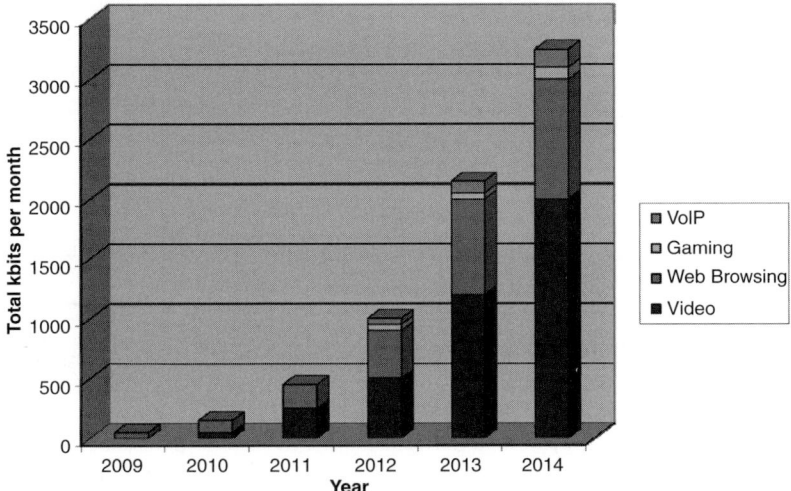

Figure 1.4. An example of the growth of mobile data usage (adapted from [1]).

and WiMAX is driven by the exponential growth in mobile broadband data usage. As shown in Figure 1.4 (adopted from [1]), mobile data usage is expected to increase by a factor of 20–40 by 2014 in total kilobits per month. This has been made possible by the advent of smart phones on the mass market and affordable broadband wireless services using laptops/ iPads/USB dongles. Hence today's networks should evolve rapidly to meet the large and rapidly growing data demand.

Table 1.2. *Video requirements for different device types/applications*

Device type	Screen size (inches)	Resolution	Average MPEG4 data rate (kbps)	Mobility	Wireless technology
Smart phones	2.5–3	QVGA (320 × 240)	240	Full	3G/4G
Multimedia phones	3–3.5	HVGA (480 × 320)	600	Full	3G/4G/ 4.5G
Personal media players	4.7	VGA (640 × 480)	900	Full	4G/4.5G
Standard-definition TV	<32	SD 480i (1280 × 720)	1500	Full	4G
Laptops	12–7	HD 720i (1280 × 720)	3500	Nomadic	4G/4.5G
Low-tier HD TV	<32	HD 720p (1280 × 720)	7000	Fixed	4G/4.5G
High-tier HD TV	>32	HD 1080p (1920 × 1080)	14000	Fixed	4G/4.5G

There seems to have been a paradigm shift in mobile data usage during the past 20 years. There is an increased demand for video data and the cellular network supporting 4G/4.5G systems should be able to cope with this increased demand. The video requirements for different types of applications/devices are shown in Table 1.2.

It may be observed from Table 1.2 that a data rate of 1–4 Mbps is required in order to support video in 4/4.5G wireless systems. Further supporting video with higher quality and low latency over wireless links requires higher bandwidths and the attributes of a 4G system such as LTE-A. The concept of a heterogeneous network (HetNet) has been introduced in LTE-A to address the capacity and coverage challenges resulting from

Table 1.3. *E-UTRA FDD operating bands [2]*

Band number	Operating frequencies (MHz)		Spectrum availability
	Downlink	Uplink	
1	1920–1980	2110–2170	EMEA, Brazil, Japan, India
2	1850–1910	1930–1990	Americas
3	1710–1785	1805–1880	EMEA, Asia–Pacific
4	1710–1755	2110–2155	USA
5	824–849	869–894	Americas, Australia
6	830–840	875–885	Japan
7	2500–2570	2620–2690	EMEA, China, S. America, Canada
8	880–915	925–960	EMEA, Asia–Pacific, S. America
9	1749.9–1784.9	1844.9–1879.9	Japan
10	1710–1770	2110–2170	Americas
11	1427.9–1447.9	1475.9–1495.9	Japan
12	698–716	728–746	USA
13	777–787	746–756	USA
14	788–798	758–768	USA
17	704–716	734–746	USA
18	815–830	860–875	Japan
19	830–845	875–890	Japan
20	832–862	791–821	EMEA
21	1447.9–1462.9	1495.9–1510.9	Japan
22	3410–3500	3510–3600	EMEA, S. America, Asia–Pacific
[23]	2000–2020	2180–2200	USA
[24]	1626.5–1660.5	1525–1559	USA
[25]	1850–1915	1930–1995	Americas

EMEA, Europe, Middle East, and Africa.

the enormous growth of data services. In the HetNet concept, the traditional macro network is deployed to provide umbrella coverage and the augmenting component provides an underlay network that could be either a new type of low-power network nodes (pico-cells, relays, and femto-cells) or a complementary technology like Wi-Fi. It will be shown in Section 6.5 how pico-cells can offer an order-of-magnitude improvement

Table 1.4. *E-UTRA TDD operating bands [2]*

Band number	Operating frequencies (MHz)	Spectrum availability
33	1900–1920	Africa, China
34	2010–2025	Africa, China
35	1850–1910	N. America
36	1930–1990	N. America
37	1910–1930	N. America
38	2570–2620	Africa, Europe, S. America
39	1880–1920	China
40	2300–2400	Asia, China, Europe
41	2496–2690	Africa, Americas, Asia, Europe
[42]	3400–3600	EMEA, Americas, Asia–Pacific
[43]	3600–3800	EMEA, S. America, Asia–Pacific

in user experience over a traditional macro-cell network and thus can support the video requirements of different devices as shown in Table 1.2.

1.3 Radio spectrum for wireless broadband

Radio spectrum for wireless broadband is available in different frequency bands and comes as both paired and unpaired bands. The radio-spectrum availability and regulations also vary among geographical areas. The 4G/4.5G wireless broadband technologies are designed to operate in different spectrum allocations using both paired (FDD) and unpaired (TDD) spectrum. The E-UTRA operating bands and spectrum availability for FDD are shown in Table 1.3 [2]. Note that the FDD operating bands 22–25 are currently under discussion in 3GPP and the final details may change.

The corresponding E-UTRA TDD operating bands and spectrum availability are shown in Table 1.4 [2]. Note that the operating TDD bands 42 and 43 are currently under discussion in 3GPP and the final details may change.

Operators and regulators across the world are trying to clear enough spectrum to deploy 4G/4.5G wireless broadband technologies based on LTE/LTE-A or WiMAX Rel-1/Rel-2 to meet the increased demand for mobile data usage which tends to account for approximately 60% of the service revenue. More spectrum is also necessary in order to provide the higher quality of service required for applications such as video, video-conferencing, and gaming.

References

[1] Morgan Stanley, *The Mobile Internet Report Setup*, December 15, 2009.
[2] 3GPP TS 36.101, UE radio transmission and reception, v8.5.0, March 2009.

Additional reading

[1] Halonen, T., Romero, J., Melero, J., *GSM, GPRS and EDGE Performance, Evolution Towards 3G/UMTS*, 2nd edition, Wiley, 2003.
[2] Iniewski, K., *Internet Networks, Wired, Wireless and Optical Technologies*, CRC Press, 2009.
[3] Andrews, J., Ghosh, A., Muhamed, R., *Fundamentals of WiMAX*, Prentice Hall, 2007.
[4] Dahlman, E., Parkvall, S., Skold, J., Beming, P., *3G Evolution, HSPA and LTE for Mobile Broadband*, 2nd edition, Academic Press, 2008.

2   LTE overview

2.1 Introduction

Long Term Evolution (LTE) of the Universal Mobile Telecommunications System (UMTS) was developed to ensure that the technology remains competitive for the foreseeable future. Requirements for the LTE Rel-8 system include improved system capacity and coverage, improved user experience through higher data rates and reduced latency, reduced deployment and operating costs, and seamless integration with existing systems. The requirements may be broken down into different categories – system performance, latency, coverage, deployment, and complexity. To achieve these goals, new designs for the radio access networks and system architectures are needed.

A representative list of LTE Rel-8 requirements for the radio access networks is given in Table 2.1 while the complete set of requirements may be found in [1]. From a system and user performance perspective, the following requirements have been defined: peak data rate, cell spectral efficiency, cell-edge user throughput, and average user throughput. For the downlink, peak data rates of at least 100 Mbps must be supported for a system bandwidth of 20 MHz, while for the uplink, peak data rates of at least 50 Mbps must be supported. Cell, cell-edge user, and average user performance requirements are defined in terms of spectral efficiency (i.e. supportable throughput in bits per second per MHz) and in relation to Rel-6 UMTS performance. In general, improvement by a factor of 3–4 is expected in the downlink while improvement by a factor of 2–3 is expected in the uplink.

Latency requirements are also defined for the control and user planes. For the user plane (U-plane), a maximum latency of 5 ms is desired. This latency is measured as the one-way delay from when a packet is available at the Internet Protocol (IP) layer to when it arrives at the User Equipment (UE).

Table 2.1. *LTE Rel-8 requirements*

Feature	Requirements
Peak data rate	Downlink – 100 Mbps at 20 MHz
	Uplink – 50 Mbps at 20 MHz
Cell spectral efficiency	Downlink – 3–4 times Rel-6 HSDPA
	Uplink – 2–3 times Rel-6 HSUPA
Cell-edge user spectral efficiency	Downlink – 2–3 times Rel-6 HSDPA
	Uplink – 2–3 times Rel-6 HSUPA
Average user spectral efficiency	Downlink – 3–4 times Rel-6 HSDPA
	Uplink – 2–3 times Rel-6 HSUPA
C-plane latency	100 ms from camped to active state
	50 ms from dormant to active state
C-plane capacity	400 users
U-plane latency	5 ms
Broadcast service	Spectrum efficiency of 1 bit/s per Hz
Mobility	Up to 350 km/h
Maximum cell range	100 km
Spectrum support	Flexible (up to 20 MHz)

Control-plane (C-plane) latencies are defined using two different requirements. The first requirement is that the transition time from camped to active state should be less than 100 ms, and the second requirement is that the transition time from dormant to active state should be less than 50 ms.

In terms of deployment, LTE must support a cell radius of up to 100 km, user speeds of up to 350 km/h (e.g. in a high-speed train environment), and flexible spectrum. However, the performance requirements given in Table 2.1 can be relaxed in more challenging deployment scenarios. For example, slight degradations are allowed for a cell radius greater than 5 km but less than 30 km. For a radius between 30 and 100 km, the requirements are further relaxed to the point that the system should be operational. LTE will also support enhanced broadcast services with the requirement of 1 bit/s per Hz throughout the coverage area, which means that, with a system bandwidth of 10 MHz, users can enjoy an aggregate downlink throughput of 10 Mbps. This is equivalent to receiving 50 simultaneous television or radio channels at a data rate of 200 kbps each.

A feasibility study was conducted in 3GPP to determine whether these requirements can be met. Results of the feasibility study were captured in [2] with the conclusion that LTE requirements can be achieved. However, enhancements in both radio and core networks are needed, including a new physical layer and a redesign of the network architecture. Enhancements to the radio access networks were made under LTE while evolution to the core network was done under System Architecture Evolution. This chapter provides a basic overview of the LTE system architecture, including a basic introduction to the new frequency-domain transmission schemes being used in the physical layer. The details and performance of the new physical layers are described in other chapters of this book.

2.2 System architecture

The LTE system architecture is based on the IP and therefore is designed to efficiently support packet-based transmission. A simplified illustration of the LTE system architecture is shown in Figure 2.1 [3]. Two main components comprise the LTE system architecture – the Evolved Universal Terrestrial Radio Access Network (E-UTRAN) and the Evolved Packet Core (EPC). The E-UTRAN is responsible for management of radio access and provides user- and control-plane support to the UEs. The user plane refers to a group of protocols used to support user data transmission throughout the network, while the control plane refers to a group of protocols for controlling the user data transmission and managing the connection between the UE and the networks. Some of these connection-management functions include handover, service establishment, resource control, etc. The E-UTRAN consists of only the eNodeBs (or eNBs, where eNB is the LTE terminology for a base station). The EPC is a mobile core network and its main responsibilities include mobility management, policy management, and security. The EPC consists of the Mobility Management Entity (MME), the Serving Gateway (S-GW), and the Packet Data Network

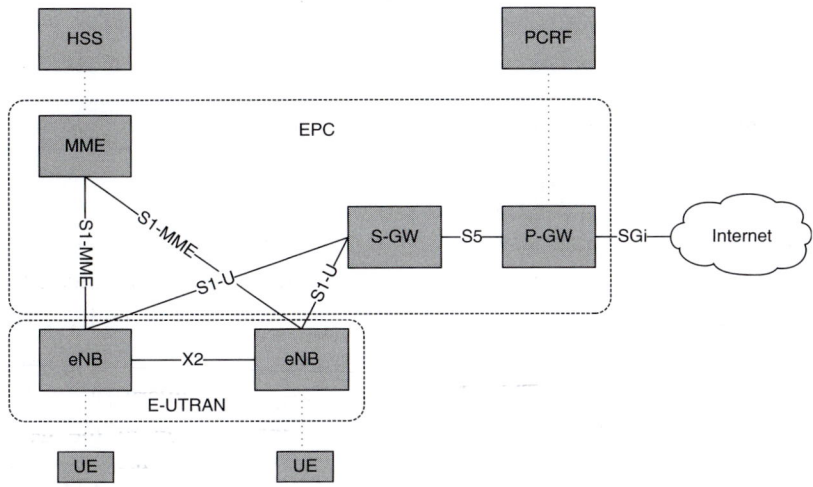

Figure 2.1. LTE system architecture.

Gateway (P-GW). Compared with previous 3GPP architectures, this new architecture has fewer nodes and therefore smaller user-plane latency [4]–[5]. This, however, requires the eNB to perform additional user-plane functions not traditionally done at the base station, such as ciphering. Both the E-UTRAN and the EPC are responsible for the quality-of-service (QoS) control in LTE [6]–[8].

Two main interfaces are defined to provide communication between different LTE entities – the S1 and X2 interfaces. The X2 interface provides communication among eNBs and can be used to transfer user- and control-plane information. Examples of possible inter-eNB exchanges using the X2 interface include handover information, measurement and interference coordination reports, load measurements, eNB configuration setups, and forwarding of user data. The S1 interface is used to connect the eNBs to the EPC (either to the MME or S-GW). The interface between eNB and S-GW is called S1-U and is used to transfer user data. The interface between eNB and MME is called S1-MME and is used to transfer control-plane information. Examples of control-plane information include mobility support, paging, data service management, location services, and network management [9].

eNB–eNB: X2   eNB–MME: S1–MME

                                 eNB–S–GW: S1–U

[handwritten: PDCP: Header compression, ciphering
RLC: Segmentation, ARQ, Err detection
and recovery]

2.2.1 E-UTRAN

The E-UTRAN provides air-interface user- and control-plane protocol management for the users. It supports the following functions: radio resource management, measurements, access-stratum security, IP header compression and encryption of the user data stream, MME selection, user-plane data routing to the S-GW, and scheduling and transmission of paging messages, broadcast information, and public warning system messages [3].

The following user-plane protocols are supported: Packet Data Convergence Protocol (PDCP), Radio Link Control (RLC), Medium Access Control (MAC), and Physical layer (PHY). An illustration of the user-plane data flow chart is shown in Figure 2.2. A radio bearer is used to transfer data and control between the UE and the E-UTRAN. User-plane data is sent via traffic radio bearers while control-plane data is sent using the signaling radio bearers. Several bearers may be established for the same user, on the basis of traffic types and characteristics. The PDCP sublayer is responsible for header compression and decompression, security functions such as ciphering and deciphering, and transfer of user protocol data units to the RLC. In addition, the PDCP sublayer also ensures in-sequence delivery of user-plane data and retransmission of service data units during handover. The RLC sublayer will perform segmentation and reassembly, ARQ error correction, and delivery of the user protocol data units to the MAC via appropriate logical channels. The RLC sublayer is also responsible for protocol-error detection and recovery, RLC re-establishment, and in-sequence delivery of the protocol data units.

The main functions of the MAC sublayer include mapping between logical and transport channels, scheduling of data, HARQ, priority handling, and multiplexing/de-multiplexing of MAC service data units. A logical channel is a data-transfer service offered by the MAC to the RLC, and can be classified as either a control or a traffic channel. Control logical channels are used to transfer control-plane information while traffic logical channels are used to transfer user-plane information. In LTE, only one logical traffic channel, the Dedicated Traffic Channel (DTCH),

[handwritten: Mac: Scheduling, HARQ, Priority handling, Mux service data
Mapping logical and transport channels]

Radio Bearers

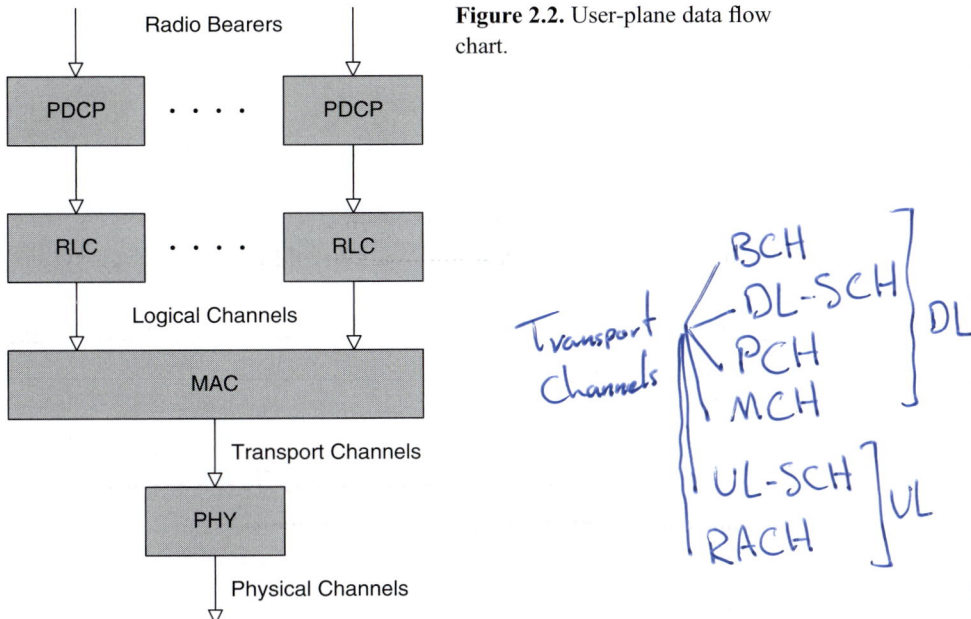

Figure 2.2. User-plane data flow chart.

is defined. However, four logical control channels are used – the Broadcast Control Channel (BCCH), Paging Control Channel (PCCH), Common Control Channel (CCCH), and Dedicated Control Channel (DCCH). The BCCH is used for broadcasting system control information while the PCCH is used for paging. The CCCH is used for control when the UE has no RRC connection with the network, while the DCCH is used to transmit dedicated control information when the UE has an RRC connection to the network.

At the MAC, data from the logical channels is multiplexed and delivered to the PHY on the transport channels. The MAC is also responsible for all the scheduling and associated functions such as selection of transport-block size, error correction using HARQ, and priority handling of the different logical channels and users. The following transport channels are available in the downlink: the Broadcast Channel (BCH), Downlink Shared Channel (DL-SCH), Paging Channel (PCH), and Multicast Channel (MCH). Uplink transport channels include the Uplink Shared Channel (UL-SCH) and the Random Access Channel (RACH).

At the PHY, the transport channels are mapped into physical channels for transmission over the air interface. The PHY is responsible for cyclic redundancy check (CRC) insertion, channel coding, HARQ processing, scrambling, modulation, link adaptation, power control, and resource mapping.

The control plane supports the Radio Resource Control (RRC) protocol. The main functions of the RRC sublayer include system information broadcast, paging, connection management, security, radio bearer management, mobility functions, QoS management functions, and UE measurement reporting. Two RRC states are used in LTE – idle and connected. In the idle state, a user has been assigned an identity but no RRC connection. In the connected state, a user has an RRC connection, which allows data transmission and reception with the network.

2.2.2 Evolved packet core

In 2G and 3G mobile broadband systems, two separate core networks were needed – circuit-switched for voice applications and packet-switched for data applications. In LTE, only a packet-switched mobile core based on the IP is used. This allows a flat architecture with a smaller number of network entities. This mobile packet core is called the Evolved Packet Core (EPC) in LTE. The EPC consists of the following network elements: the Mobility Management Entity (MME), Serving Gateway (S-GW), and Packet Data Network Gateway (P-GW). In addition, two additional logical network elements – the Home Subscriber Server (HSS) and the Policy Control and Charging Rules Functions (PCRF) – are also generally included as part of the core network. In LTE, the EPC provides the following functions: mobility management, session management, security management, and policy control and charging. Mobility management provides signaling support between the UE and the network using the Non-Access Stratum (NAS) protocols. Session management refers to the establishment and management of data bearers. Security management provides data encryption and authentication services for the users. Policy control and charging refers to access and control of services as prescribed by the operator including QoS management, metering, service control

EPC ⟨ MME, S-GW, P-GW ⟩ Main elements ⟨ HSS, PCRF ⟩

based on user classification, and policy control enforcement. It also is responsible for charging and billing of services.

The MME serves as the control entity for the EPC and provides the following main functionalities: NAS signaling and security, P-GW and S-GW selection, roaming support, user authentication, bearer management, and idle-state mobility handling. The NAS is a functional layer that provides signaling and traffic between the UE and the packet data network gateway.

The S-GW manages the user data plane between the eNBs and the packet data network gateway. As UEs move across areas served by different eNBs, the S-GW serves as a mobility anchor ensuring continuous data connection. This includes mobility management for handovers between LTE and other 3GPP technologies. The S-GW is connected to the eNBs via the S1-U interface.

The P-GW provides data connectivity to the external packet data networks such as the Internet or IP Multimedia Subsystem (IMS) networks. IMS networks are used to provide multimedia services such as Voice over Internet Protocol (VoIP), video conferencing, and messaging. Functions of the P-GW include packet filtering and routing, IP address allocation, charging and policy enforcement via the PCRF, and lawful interception.

The HSS maintains a database of subscriber-related information [10]. This includes user profile and state information including restrictions on roaming, QoS, access-point information, security information, location information, and access/service authorization. It also stores information about available P-GWs that a user can connect to.

The PCRF has two main functions – policy control and flow-based charging [11]. It defines the rules and policy associated with this work. For example, different charging rules such as volume-based, time-based, and event-based can be enforced. In addition, different rates based on user characteristics (e.g. roaming versus home) and service characteristics (e.g. traffic type or guaranteed data rate) can be applied. Policy control functions include gating control, QoS control, and usage monitoring. Gating refers to blocking of packets, while usage monitoring refers to monitoring of network resources. In LTE, all services are delivered through IP-based connections. To allow differentiation between different

Table 2.2. *QoS class-identifier (QCI) parameters [11]*

QCI	Minimum guaranteed bit rate	Scheduling priority	Packet delay budget (ms)	Packet error loss rate
1	Yes	2	100	10^{-2}
2	Yes	4	150	10^{-3}
3	Yes	5	300	10^{-6}
4	Yes	3	50	10^{-3}
5	No	1	100	10^{-3}
6	No	7	100	10^{-3}
7	No	6	300	10^{-6}
8	No	8	300	10^{-6}
9	No	9	300	10^{-6}

traffic services, the PCRF network element can provide different QoS for different data-bearer classes. Each data-bearer class is associated with a QoS class identifier (QCI) that specifies the following attributes: whether a minimum bit rate is guaranteed, scheduling priority, packet delay budget, and packet loss rate. Nine different QCI profiles are defined in [11] and shown in Table 2.2. Note that a user can be configured with many simultaneous data bearers (e.g. one for VoIP and one for TCP-based services), and each data bearer will be assigned its own QCI.

Different QCI values may be assigned to different service types on the basis of service characteristics. For example, VoIP is usually assigned a QCI value of 1 since it requires a minimum guaranteed bit rate, has a low packet delay budget, and is somewhat tolerant of packet losses. TCP-based services such as web browsing and email are usually assigned a QCI value of 8 since these services do not require a guaranteed bit rate and are not delay-sensitive but must have very low packet losses.

2.2.3 User equipment

In LTE, the user equipment communicates with the E-UTRAN and EPC using relevant radio protocols. User-plane communication terminates at

Table 2.3. *User equipment category for LTE Rel-8 [12]*

| UE category | Maximum number of bits in a subframe | | Maximum number of downlink MIMO layers |
	Downlink	Uplink	
1	10296	5160	1
2	51024	25456	2
3	102048	51024	2
4	150752	51024	2
5	299552	75376	4

the eNB and supports all user-plane protocols described in Section 2.2.1. On the control-plane side, UE communicates with the eNB via the RRC protocol, and with the MME via the NAS control protocol.

To support different hardware capabilities, different user equipment categories are defined as shown in Table 2.3 [12]. The categories are distinguished through the maximum supported data rates for downlink and uplink. In addition, the maximum number of data layers (or streams) may differ depending on UE category. For example, the maximum down-link and uplink data rates for UE category 1 are approximately 10.3 and 5.2 Mbps, respectively. In addition, this UE category does not support reception of more than one downlink data stream simultaneously. On the other hand, UE category 5 can support four downlink data streams simultaneously. In addition, it is capable of transmitting using 64-QAM modulation in the uplink. As a result, the maximum downlink and uplink data rates for this UE category are approximately 299.6 and 75.4 Mbps, respectively. The UE category is transmitted by the UE to the E-UTRAN during call setup using the RRC protocol.

In addition to the UE category, different UE capabilities are defined separately via the feature group indicators. These feature group indicators are transmitted by the UE to the network as part of the call setup procedure and are used to inform the E-UTRAN of the UE capabilities with respect to certain predefined LTE features. They include, for example, whether the UE can support inter-frequency handover, periodic measurements for self-optimized networks, inter-radio access technology measurements, intra-subframe frequency hopping in the uplink, simultaneous transmission of uplink control information, and semi-persistent scheduling. On the basis of the reported UE category and capabilities, the E-UTRAN can be aware of the different features that can be supported by the user.

2.3 Transmission scheme

In the downlink, OFDMA has been selected as the transmission technique for LTE. In OFDMA, the frequency resource is divided into parallel-frequency subcarriers [13]–[14]. Each subcarrier is capable of carrying one modulation symbol. Different subcarriers are grouped together to form a sub-channel that serves as the basic unit of data transmission. The main reasons why OFDMA was selected as the basic transmission scheme for LTE are its high spectral efficiency, low-complexity implementation, and the ability to easily support advanced features such as frequency-selective scheduling, multiple-input multiple-output (MIMO) transmission, and interference coordination [15]. In the uplink, SC-FDMA was selected due to its ability to provide similar advantages to OFDM such as orthogonality among users, frequency-domain scheduling, and robustness with respect to multipath operation. However, SC-FDMA has a lower requirement for low-power amplifier back-off or de-rating. As a result, the average transmission power can be much higher using SC-FDMA than it can with OFDMA. This increases coverage in the uplink and provides higher uplink data rates to users at the cell edge. Comprehensive reviews of the OFDMA and SC-FDMA literature may be found in [16]–[17].

2.3.1 OFDMA

OFDMA has several advantages over the wideband code-division multiple-access (WCDMA) technique used in the previous generations of UMTS. As demonstrated in [2], OFDMA provides better performance in terms of spectral efficiency (i.e. how much data can be transmitted for a given amount of bandwidth) than does WCDMA both for broadcast and for unicast services. This is due to the lack of inter-symbol interference

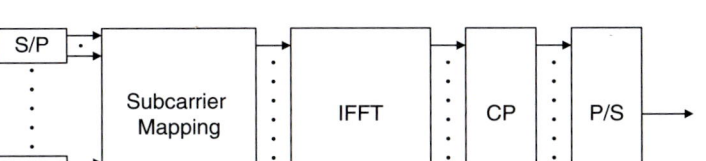

Figure 2.3. Block diagram for OFDMA.

from multipath channels and the absence of intra-cell interference because users are orthogonal (i.e. they do not interfere with each other) in the frequency domain. In addition, the OFDMA transmission technique scales easily to different bandwidths, so multiple system bandwidth configurations can be efficiently supported. In addition, low-complexity receivers can be used with OFDMA.

In addition, frequency-domain scheduling and MIMO processing techniques can be used. An example of frequency-domain scheduling techniques is frequency-selective scheduling. In frequency-selective scheduling, users are assigned data only on good frequency bands (i.e. bands with large gain), which are determined on the basis of channel quality feedback from the UE. For broadcast services, single-frequency broadcast networks can be supported. In this case, multiple base stations transmit the same broadcast signals. The signals are coherently combined at the user, thus improving performance at the cell edge substantially.

A basic block diagram illustrating OFDMA signal generation for one OFDM symbol is shown in Figure 2.3. Data symbols from different users are mapped to different subcarriers depending on the frequency bands assigned to those users. This is done in the frequency domain. The information is then subjected to an inverse fast Fourier transform (IFFT) to convert the frequency-domain subcarriers into time-domain signals. A cyclic prefix is then added, and the signal is ready for transmission. Note that the basic transmission unit for data is a subframe that spans multiple OFDM symbols. At the receiver, the reverse operation is performed. The cyclic prefix is removed, then the time-domain signal is subjected to a fast Fourier transform (FFT) so that the modulation symbols on each subcarrier can be extracted. Each user then extracts the frequency

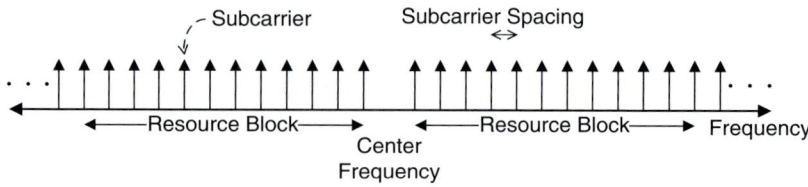

Figure 2.4. Frequency-domain illustration of OFDM.

resource units corresponding to his assigned subcarriers. Equalization is performed and the data is passed onward for decoding.

A frequency-domain illustration of OFDM transmission is shown in Figure 2.4, where each data symbol is modulated onto one of the subcarriers. The OFDM parameters must be selected carefully in order to meet LTE requirements while minimizing overhead. Key design parameters include cyclic-prefix length, subcarrier spacing, and resource-block size. In LTE, the direct-current (DC) subcarrier (the subcarrier at the center frequency) is not used since the performance of this subcarrier can be very poor for certain transmitter and receiver designs. Thus, the usable subcarriers are located around this center frequency as shown in Figure 2.4. The subcarrier spacing is the frequency spacing between two adjacent subcarriers. Small subcarrier spacing means that more subcarriers are available for a given amount of bandwidth, thus increasing the spectral efficiency since more data symbols are available for a given amount of bandwidth. In addition, small subcarrier spacing also ensures that the fading on each subcarrier is frequency-non-selective. However, performance degrades as subcarrier spacing decreases due to Doppler shift and phase noise. Doppler shift is caused by UE movement with larger shift as UE velocity increases. This causes inter-carrier interference whose degradation increases as the subcarrier spacing decreases. Phase noise is caused by fluctuations in the frequency of the local oscillator, and will cause inter-carrier interference as well. To minimize performance degradation from phase noise, the subcarrier spacing should be greater than 10 kHz. Furthermore, to support UE up to a speed of 350 km/h, the subcarrier spacing should be around 9–17 kHz. As a result, a subcarrier spacing of 15 kHz was chosen for LTE.

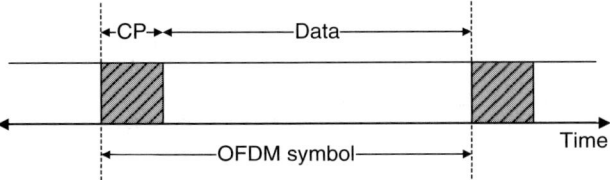

Figure 2.5. OFDM symbol in the time domain.

In LTE, frequency resource is assigned in units of resource blocks. Several factors must be considered in the selection of the resource block size in frequency. First, it should be small enough that the frequency-selective scheduling (i.e. scheduling data transmission on good-frequency subcarriers) gain is large. Small resource-block size ensures that the frequency response within each resource block is similar, thus enabling the scheduler to assign only good resource blocks. However, since the eNB does not know which resource blocks are experiencing good channel conditions, the UE must report this information back to the eNB. Thus, the resource-block size must be sufficiently large that the feedback overhead is not too high. It also should be sufficiently large to minimize downlink control signaling, which must be used to inform the UE of its resource allocation. In [18], performance analysis of frequency-selective scheduling was performed. It was found that a resource block of size 200–900 kHz provides good performance. Since, in LTE, a subframe size of 1 ms is used to ensure low latency, the resource block size in frequency should be small so that small data packets can be efficiently supported. As a result, 180 kHz (12 subcarriers) was chosen as the resource-block bandwidth.

A cyclic prefix is needed for OFDMA transmission in order to prevent inter-symbol interference from previously transmitted OFDM symbols. The OFDM symbol with cyclic prefix and data is shown in Figure 2.5. Note that the cyclic prefix does not carry useful data and is removed at the receiver prior to processing. As a result, it is desirable to have as small a cyclic prefix as possible in order to minimize the overhead. In general, the length is chosen on the basis of the expected delay spread of the propagation channel plus some margin to allow for imperfect timing alignment.

In LTE, three different cyclic-prefix values are supported – normal (~4.7 µs) and extended (16.6 µs) for subcarrier spacing 15 kHz and extended (33 µs) for subcarrier spacing 7.5 kHz. Note that the subcarrier spacing 7.5 kHz can be used only for broadcast transmission. The normal cyclic-prefix length is approximately 4.7 µs and is sufficient to handle channel delay spread in most urban and suburban environments. With the data portion of the OFDM symbol occupying approximately 66.7 µs, this represents a cyclic-prefix overhead of 7%. An extended cyclic prefix of length 16.7 µs can be used for environments with longer delay spread and for broadcast services. In this case, however, a cyclic-prefix overhead of 25% is incurred.

To support different system bandwidths, different FFT sizes are used in order to keep the OFDM symbol duration constant. That is, regardless of the bandwidth, each OFDM symbol is of duration 66.7 µs. This allows the same subcarrier separation to be supported, thus ensuring that the same frequency-domain techniques can be applied across multiple bandwidths. Keeping the symbol duration constant also results in the same subframe length for all bandwidths, which is very attractive from a design perspective. Although the generation of the OFDM signal is up to implementation, as a guideline an FFT size of 2048 is used for 20 MHz. The FFT sizes for other bandwidths are then scaled from this value. For example, for 5 MHz, an FFT size of 512 can be used.

The eNB transmission characteristics have to satisfy the standards set by the 3GPP Radio Access Network Working Group 4 (RAN4) which is in charge of radio performance and protocol aspects. They include maximum transmit power and dynamic range, unwanted emission, frequency error, and error-vector magnitude (EVM). The frequency error is defined as the difference between the actual transmit frequency and the desired frequency. The frequency error is measured over one transmission subframe and must be within ±0.05 ppm to pass. The EVM is a measurement of the difference between the transmit signal waveform and its idealized counterpart. The higher the EVM, the worse the performance. In LTE, the EVM must be below 17.5%, 12.5%, and 8% for QPSK, 16-QAM, and 64-QAM modulation, respectively. The reason why the EVM requirement is higher for lower-order modulation is that transmission using lower-order modulation can tolerate more signal distortion before performance loss is substantial. Unwanted emissions are emissions that occur outside of the occupied bandwidth, thus generating interference with other radio systems using nearby spectrum. Unwanted emissions can be caused by the modulation process (called out-of-band emissions) or by imperfections in the transmitter (called spurious emissions).

2.3.2 SC-FDMA

In the uplink, SC-FDMA is selected due to its ability to provide similar advantages to OFDM, such as orthogonality among users, frequency-domain equalization, and robustness with respect to multipath operation while maintaining a low power amplifier back-off or de-rating requirement [19]. The key characteristic of single-carrier transmission is that each data symbol is transmitted using the entire allocated bandwidth. This is different than OFDM, where each data symbol is transmitted using only one subcarrier. Since single-carrier transmission spreads the data power over the entire bandwidth, it requires lower power amplifier back-off. The power back-off is the required reduction in the mean transmission power to ensure that the maximum power stays within the linear region of the power amplifier. Operating outside of the linear region of the power amplifier causes signal distortion and interference. For instance, given the maximum transmit power of 23 dBm (equivalent to 200 mW) and a power amplifier back-off requirement of 3.4 dB for an OFDM signal, the maximum mean transmission power is reduced to 19.6 dBm, which will reduce uplink coverage significantly. A good measure of the power back-off requirement is the cubic metric, defined in [20] as the cubic power of the signal of interest compared with a reference signal. Table 2.4 provides the cubic-metric values for OFDMA and SC-FDMA. Another measure of the power back-off requirement is the peak-to-average power ratio (PAPR). A PAPR comparison between OFDMA and SC-FDMA has been presented in [21], showing that the results for SC-FDMA are similar to those for the cubic-metric gain shown in Table 2.4. The PAPR, however, has been shown to be a less accurate predictor of amplifier power back-off than the cubic metric [22].

Table 2.4. *Comparison of cubic metric between OFDMA and SC-FDMA*

	Cubic metric	
Modulation	OFDMA	SC-FDMA
QPSK	3.4	1.0
16-QAM	3.4	1.8
64-QAM	3.4	2.0

From Table 2.4, it can be seen that SC-FDMA has a significantly lower cubic metric than that for OFDMA. For cell-edge users, where QPSK modulation is generally used, SC-FDMA enjoys a cubic-metric advantage of 2.4 dB over OFDMA. This means that cell-edge users can transmit at 1.74 times higher average power with SC-FDMA than with OFDMA for the same maximum-power limitation. As a result, for the same uplink cell-edge data rate, SC-FDMA can provide greater coverage. For example, in [21] it was shown that SC-FDMA can provide 20% greater range at a data rate of 1 Mbit/s. Conversely, for the same coverage area, SC-FDMA can deliver a higher data rate from the cell-edge users. For example, at a distance of 0.8 km from the cell, SC-FDMA can deliver a data rate of 200 kbit/s, compared with 70 kbit/s for OFDMA. This is the primary reason why SC-FDMA is selected for the uplink. The low power back-off property is accomplished by transmitting the data symbols serially rather than in parallel like in OFDMA, which results in substantially reduced signal fluctuations. This helps conserve battery life or extend the range by reducing the back-off due to non-linearity in the power amplifier. The performance of SC-FDMA, however, is not as good as that of OFDMA given the same type of receiver. The performance for QPSK modulation is approximately the same, while OFDMA outperforms SC-FDMA by 0.5–1 dB for 16-QAM [21]. Although this negates the benefits of SC-FDMA somewhat, especially for indoor users, coverage and cell-edge data rate were seen as the most important criteria in the uplink.

In LTE, discrete Fourier transform–spread–OFDM (DFT-S-OFDM) is used to generate the SC-FDMA signal in the frequency domain as shown in Figure 2.6 [23]–[24]. Note that generation of the SC-FDMA signal using

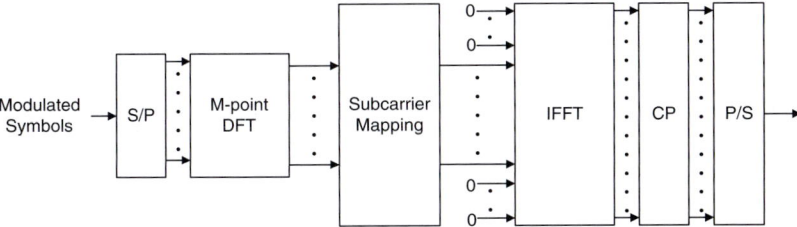

Figure 2.6. Block diagram for SC-FDMA.

DFT-S-OFDM is almost identical to that of OFDM, with the exception of the additional M-point discrete Fourier transform (DFT). Although DFT processing is more computationally intensive than the FFT, efficient implementation for certain DFT sizes is available. Specifically, DFTs of prime length can be calculated using efficient FFT algorithms. The method shown in Figure 2.6 generates SC-FDMA signal in the frequency domain. This allows frequency-domain pulse shaping to be applied prior to the IFFT to further reduce the cubic metric [25].

The first M-point DFT is used to provide frequency-domain precoding, which is mapped to M contiguous-frequency subcarriers prior to the IFFT. To preserve the single-carrier property, transmission from a user within an SC-FDMA symbol must be either contiguous or evenly spaced in the frequency domain.

Two different types of single-carrier transmission can be generated using DFT-S-OFDM, depending on how the resource-element mapping is done. The mapping may be done such that a distributed or localized frequency allocation is generated as shown in Figure 2.7. Localized mapping means that the entire allocation is contiguous in frequency. This allows good channel-estimation performance since the pilots are contiguous, thus interpolating techniques can be used in channel estimation. In addition, it will be easy to multiplex different users together in the spectrum. However, frequency diversity is poor. Distributed mapping means that the allocated bandwidth is evenly distributed in frequency. This provides very good frequency diversity. However, the pilots must be distributed, and thus channel-estimation performance suffers. It can also be difficult to multiplex all the users together in the spectrum. In addition,

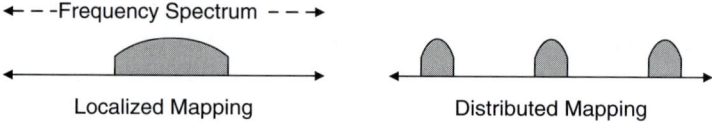

Figure 2.7. Localized versus distributed mapping for DFT-S-OFDMA.

frequency-selective scheduling where a user is assigned only a selected portion of the spectrum (generally one that is providing good radio conditions) cannot be taken advantage of.

Performance comparisons of localized versus distributed mapping using realistic channel estimation have been published in [22]. The results showed that the two methods provide similar performance. The gain in frequency diversity from distributed transmission is lost through poorer channel-estimation performance. Given these performance results and other difficulties with scheduling of users, only localized mapping is supported in LTE. However, to provide frequency diversity, hopping, whereby the user hops from one localized frequency assignment to a different frequency, can be used.

At the receiver, the reverse operation of the transmitter functions is performed for data demodulation. The received signal first undergoes RF processing and analog-to-digital conversion. Then the cyclic prefix is removed and an FFT is performed. Channel estimation is performed on the basis of the pilots that have been embedded into the transmission packet. In addition to channel estimation, frequency and timing estimation and correction may also be performed. Subcarrier demapping and equalization is done next, followed by an IDFT and finally an M-point IDFT. Unlike in conventional FDMA, the addition of an M-point DFT/IDFT is used to spread out each modulated data symbol onto all of the subcarriers used. This lowers the peak-to-average power of the transmission signal, resulting in higher maximum transmission power. However, because of the M-point IDFT, all the transmitted modulated symbols within the SC-FDMA symbol have the same SINR. The performance of the receiver depends on the type of receivers as well as channel estimation, frequency and time tracking, and decoding algorithms. Several types of receivers can be used for SC-FDMA, including, for example, zero-forcing receivers, minimum-mean-squared-error receivers, interference-rejection combining receivers, and turbo equalization receivers. Naturally, receiver performance is tied to complexity, with performance improving as complexity grows. In practice, a minimum-mean-squared-error or interference-rejection combining receiver is usually used because of its good performance and manageable complexity.

In the uplink, the same issue with a DC subcarrier is present as in the downlink. However, if the DC carrier is skipped in the uplink, then the cubic metric can increase if the UE transmits on the resource block that contains the DC subcarrier. This increase in cubic metric is on the order of 0.5–0.7 dB, which will reduce the maximum transmit power by 12%–18%. As a result, a carrier shift of 7.5 kHz was introduced in the uplink so that no carrier will be centered around DC. This leads to only small performance degradation on the resource block that contains the center frequency.

Transmitter characteristics and requirements for the UE can be found in [26], including operating bands and radio-channel arrangement, transmit power, dynamic range, transmit signal quality, and output spectrum emissions. Currently, only one UE power class is defined, with a maximum output power of 23 dBm and a tolerance of ±2 dB. The minimum power output is −40 dBm. The actual UE transmit power, however, is determined using a power-control formula and controlled by the eNB. To allow for implementation margin, a tolerance band is provided. For instance, if the UE has not had an uplink transmission within the last 20 ms, a tolerance of ±9 dB is allowed. That is, the actual transmit power of the UE can be within 9 dB of the desired power. For UEs with uplink transmission within the past 20 ms, a much smaller tolerance is allowed. This tolerance is based on the difference between the required power and the last transmit power. The larger the power difference, the larger the allowable tolerance. In this case, the minimum tolerance band is ±2.5 dB for a power difference of less than 2 dB, and the maximum tolerance band is ±6 dB for a power difference of greater than or equal to 15 dB. The UE also has the same EVM requirements for QPSK (17.5%) and 16-QAM (12.5%) as the eNB. However, the EVM requirement for 64-QAM has not been defined. Also similarly to the eNB, the UE has requirements on out-of-band and spurious emissions that are intended to limit the amount of interference with other UEs. The out-of-band emissions are limited by the spectrum mask and adjacent-channel leakage ratio which the UE must satisfy. The spectrum-mask requirement limits the maximum out-of-band power level, while the requirement regarding the adjacent-channel leakage ratio limits the mean power on adjacent channel frequencies. Limits on spurious emissions are also provided in order to regulate how much interference can be generated by unwanted transmitter effects at the UE.

References

[1] 3GPP TS 25.913, Requirements for Evolved UTRA (E-UTRA) and Evolved UTRAN (E-UTRAN), v7.3.0, March 2006.
[2] 3GPP TS 25.912, Feasibility study for evolved Universal Terrestrial Radio Access (UTRA) and Universal Terrestrial Radio Access Network (UTRAN), v7.2.0, July 2006.
[3] 3GPP TS 36.300, E-UTRA and E-UTRAN overall description, v8.12.0, March 2010.
[4] Larmo, A., Lindstrom, M., Meyer, M. *et al.*, "The LTE link-layer design," *IEEE Communications Magazine*, vol. 47, no. 4, pp. 52–59, April 2009.
[5] Dahlman, E., Ekstrom, H., Furuskar, A. *et al.*, "The 3G Long-Term Evolution – radio interface concepts and performance evaluation," *IEEE 63rd Vehicular Technology Conference*, vol. 1, pp. 137–141, May 2006.
[6] Ekstrom, H., "QoS control in the 3GPP evolved packet system," *IEEE Communications Magazine*, vol. 47, no. 2, pp. 76–83, February 2009.
[7] Racz, A., Temesvary, A., Reider, N., "Handover performance in 3GPP Long Term Evolution (LTE) systems," *16th IST Mobile and Wireless Communications Summit*, July 2007.
[8] Bajzik, L., Horvath, P., Korossy, L., Vulkan, C., "Impact of intra-LTE handover with forwarding on the user connections," *16th IST Mobile and Wireless Communications Summit*, July 2007.
[9] Meng, W., Georgiades, M., Tafazolli, R., "Signaling cost evaluation of mobility management schemes for different core network architectural

arrangements in 3GPP LTE/SAE," *IEEE Vehicular Technology Conference*, pp. 2253–2258, May 2008.

[10] 3GPP TS 23.002, Network architecture, v10.0.0, September 2010.

[11] 3GPP TS 23.203, Policy and charging control architecture, v10.1.0, September 2010.

[12] 3GPP TS 36.306, User equipment (UE) radio access capabilities, v8.7.0, June 2010.

[13] Weinstein, S., Ebert, P., "Data transmission by frequency-division multiplexing using the discrete fourier transform," *IEEE Transactions on Communication Technology*, vol. 19, no. 5, pp. 628–634, October 1971.

[14] Bingham, J. A. C., "Multicarrier modulation for data transmission: an idea whose time has come," *IEEE Communications Magazine*, vol. 28, no. 5, pp. 5–14, May 1990.

[15] Sari, H., Karam, G., Jeanclaude, I., "An analysis of orthogonal frequency-division multiplexing for mobile radio applications," *IEEE 44th Vehicular Technology Conference*, vol. 3, pp. 1635–1639, June 1994.

[16] Taewon, H., Chenyang, Y., Gang, W., Shaoqian, L., Ye Li, G., "OFDM and its wireless applications: a survey," *IEEE Transactions on Vehicular Technology*, vol. 58, no. 4, pp. 1673–1694, May 2009.

[17] Ciochina, C., Sari, H., "A review of OFDMA and single-carrier FDMA," *European Wireless Conference (EW)*, pp. 706–710, April 2010.

[18] R1-050720, "Frequency selective scheduling resource block size for EUTRA downlink," Motorola, RAN1#42, San Diego, CA, October 2005.

[19] Myung, H. G., Lim, J., Goodman, D. J., "Single carrier FDMA for uplink wireless transmission," *IEEE Vehicular Technology Magazine*, vol. 1, no. 3, pp. 30–38, September 2006.

[20] R1-060385, "Cubic metric in 3GPP-LTE", Motorola, RAN1#44, Denver, CO, February 2006.

[21] R1-051088, "Coverage comparison between UL OFDMA and SC-FDMA," Nokia, RAN1#42, San Diego, CA, October 2005.

[22] R1-051033, "Further topics on uplink DFT-S-OFDM for E-UTRA," Motorola, RAN1#42, San Diego, CA, October 2005.

[23] Fan, X., Li, Y., Li, M., Zhang, X., "Analysis and comparison of different SC-FDMA schemes for 3GPP LTE," *International Conference on Wireless Communications, Networking and Mobile Computing*, pp. 787–790, September 2007.

[24] Priyanto, B. E., Codina, H., Rene, S., Sorenson, T. B., Mogensen, P., "Initial performance evaluation of DFT-spread OFDM based SC-FDMA for UTRA LTE uplink," *IEEE Vehicular Technology Conference*, April 2007.

[25] Mauritz, O., Popovic, B. M., "Optimum family of spectrum-shaping functions for PAPR reduction of DFT-spread OFDM signals," *IEEE 64th Vehicular Technology Conference*, September 2006.

[26] 3GPP TS 36.101, User equipment (UE) radio transmission and reception, v8.4.0, December 2008.

3 Downlink transmission and system performance

3.1 Introduction

In this chapter the details of LTE downlink transmission are discussed. The LTE downlink air interface uses the OFDM multiple-access technique described in Chapter 2. The use of OFDM transmission technology provides significant advantages over other radio transmission techniques. They include high spectral efficiency, support for broadband data transmission, the absence of intra-cell interference (i.e. multiple users in the same cell can share the same subframe without interfering with each other), resistance to inter-symbol interference arising from multipath operation, natural support for MIMO schemes, a low-complexity receiver, and support for frequency-domain techniques such as frequency-selective scheduling, a single-frequency network, and soft fractional frequency reuse. In addition to OFDM, LTE also utilizes several other features to enhance system performance and user experience. They include short frame size to minimize latency, a single-frequency network to provide high-data-rate broadcast services, VoIP support to increase voice capacity, coverage for very large cells, and coverage for high-speed users (up to 350 km/h) [1]–[2].

In LTE, both frequency-division duplex (FDD) and time-division duplex (TDD) transmission are supported. Furthermore, the baseband structure is common between FDD and TDD thus making LTE a very flexible and efficient technology that can be deployed in either paired or unpaired spectrum. In LTE, the differences between TDD and FDD are mostly at the physical layer. As a result, identical network architecture can be used to support both modes, thus reducing deployment costs and complexity substantially. In the specifications, the two modes have been designed to share as much functionality and as many features as possible, with the main design differences being the need to support various TDD DL/UL allocations and provide co-existence with other TDD systems.

This chapter will cover an overview of E-UTRA downlink structure, downlink reference signal structure, downlink control and other overhead channels, and Multimedia Broadcast Multicast services. Aspects of the performance of downlink shared and control channels for the single-input multiple-output case are also included. Finally, a brief comparison between LTE FDD and TDD systems is provided.

3.2 Mapping between transport and physical channels

The LTE physical layer provides four services to the MAC layer, namely data-transfer services, signaling of Hybrid Automatic Repeat Request (HARQ) feedback, signaling of scheduling request and measurement reports (e.g. channel-quality information, received signal reference power). The access to the data-transfer services is through the use of transport channels. Six physical and four transport channels are defined for LTE. The physical channels are the Physical Downlink Shared Channel (PDSCH), Physical Broadcast Channel (PBCH), Physical Multicast Channel (PMCH), Physical Control Format Indicator Channel (PCFICH), Physical Downlink Control Channel (PDCCH), and Physical HARQ Indicator Channel (PHICH). The four transport channels are the Broadcast Channel (BCH), Paging Channel (PCH), Downlink Shared Channel (DL-SCH), and Multicast Channel (MCH).

The PDSCH carries downlink data traffic and the PDSCH resources are assigned using the downlink scheduling assignment carried on the PDCCH. The PDSCH resources assigned to each user are orthogonal (i.e. they are unique in time and frequency) except for in multi-user MIMO (MU-MIMO), where multiple users are assigned overlapping time–frequency resources. The PDCCH carries control information for uplink and downlink as well as uplink power-control information, and is assigned on a per-user basis, i.e. each control channel is individually coded with its own error-detection information and can be power controlled. In addition, the PDCCH also carries uplink power-control information. The function of the PCFICH is to dynamically indicate the number of control symbols used in a subframe. This feature is useful in improving system capacity since the control-channel overhead can be reduced on the basis of user loading in a cell. The PHICH is used to carry acknowledgments associated with uplink data transmission. Finally, the BCH is used to convey basic physical-layer system parameters necessary to initiate access to the system.

The downlink also supports several types of signal, including the synchronization signal, the cell-specific reference signal, and UE-specific dedicated reference signals. The synchronization signal is used for cell acquisition, detection of frame timing, and handover purposes. The cell-specific reference signal is used for demodulation of downlink data transmitted using open- and closed-loop spatial multiplexing schemes, for estimating channel quality, and also for handoff measurements. The UE-specific reference signal is used for demodulating downlink data transmitted using beamforming, where the channel response is determined using a UE-specific reference signal. Table 3.1 summarizes the different types of downlink physical channels and signals.

3.3 LTE downlink frame structure

In the downlink, OFDM is the multiple-access technique for E-UTRA FDD and TDD systems. OFDM is well suited for high-data-rate systems operating in multipath environments because of its robustness with respect to delay spread. Multipath channels provide the potential to improve the system performance by exploiting both the frequency diversity and the frequency selectivity of the channel. In OFDM, one data symbol is transmitted per frequency subcarrier (called a resource element in LTE). This exploits the frequency selectivity of the multipath channel since the different symbols of the data packet are transmitted in different frequency locations. In addition, because only one data symbol is transmitted per resource element, low-complexity receivers can be used. This allows frequency-selective operation in addition to frequency-diverse scheduling and one-cell reuse of the available bandwidth. Furthermore, due to its frequency-domain nature, OFDM enables flexible-bandwidth operation with low complexity. Smart antenna technologies are also

Table 3.1. *Summary of downlink channels and signals*

Name	Description
Physical channels	
Physical Downlink Shared Channel (PDSCH)	Shared data channel used for transmission of scheduled user data
Physical Downlink Control Channel (PDCCH)	Control channel used for transmission of control information, including scheduling grants and power-control information
Physical Control Format Indicator Channel (PCFICH)	Used to dynamically indicate the number of OFDM symbols allocated for control signaling in a subframe
Physical HARQ Indicator Channel (PHICH)	Used to transmit downlink ACK/NACK associated with uplink data transmission.
Physical Broadcast Control Channel (PBCH)	Used to transmit system information necessary for system access
Paging Control Channel (PCH)	Used to transmit paging information
Physical signals	
Cell-specific reference signal	Common reference signal used for demodulation of data and control information, deriving CQI/PMI information, and handoff measurements
UE-specific reference signal	Dedicated reference signal dedicated to a UE mainly used for RS-based beamforming
Primary and secondary synchronization signals (PSS and SSS)	Used for a variety of purposes, including detection of frame timing, cell acquisition, computing number of antennas in BCH, and for handover purposes

easier to support with OFDM, since each subcarrier becomes flat-faded and the antenna weights can be optimized on a per-subcarrier (or block of subcarriers) basis. Another significant advantage of OFDM is with respect to broadcast services using a synchronized single-frequency network with appropriate cyclic-prefix design. Single-frequency network transmission refers to simulcast transmission of the same content from multiple cells, which, when combined over the air, significantly increases the received signal power and supportable data rates for broadcast services.

The downlink physical-layer resource grid is shown in Figure 3.1. The total downlink resource grid has a size of $N_{RB}^{DL} N_{sc}^{RB}$ subcarriers by N_{symb}^{DL} OFDM symbols, where N_{RB}^{DL} is the number of available downlink resource blocks and N_{sc}^{RB} is the number of subcarriers per resource block. A resource element is equivalent to a subcarrier and can be identified by the slot number and index (k, l), where k is the subcarrier index and l is the OFDM symbol index.

Table 3.2 provides the downlink subframe parameters for different spectrum allocations with subcarrier spacing 15 kHz. For MBMS-dedicated cells, a subcarrier spacing of 7.5 kHz is also supported for very

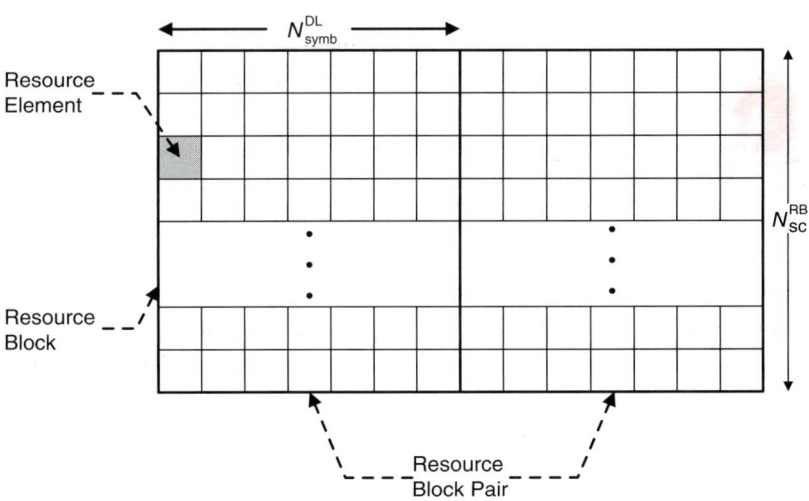

Figure 3.1. Resource grid.

Table 3.2. *Parameters for downlink transmission scheme*

	Transmission bandwidth (MHz)					
	1.4	3	5	10	15	20
Subframe duration (ms)	1.0	1.0	1.0	1.0	1.0	1.0
Subcarrier spacing (kHz)	15	15	15	15	15	15
Sampling frequency (MHz)	1.92	3.84	7.68	15.36	23.04	30.72
FFT size	128	256	512	1024	1536	2048
Number of resource blocks	6	15	25	50	75	100
Number of OFDM symbols per subframe	14/12 (normal/extended cyclic prefix)					
CP length						
Short	5.2 µs for $l = 0$, 4.7 µs for $l = 1, \ldots, 6$					
Extended	16.7 µs					

large cells. Three different cyclic-prefix values are supported – normal (~4.7 µs) and extended (16.6 µs) for subcarrier spacing 15 kHz and extended (33 µs) for subcarrier spacing 7.5 kHz. In OFDM, a cyclic prefix is used to guard against inter-symbol interference from previous OFDM symbols. Inter-symbol interference degrades performance since the received signal of the current symbol is distorted by signals from previous symbols, and complicated signal-processing techniques are required in order to mitigate this effect. In general, the longer the cyclic prefix, the more robust the protection against inter-symbol interference. However, a cyclic prefix is considered overhead since useful data cannot be carried on the cyclic prefix. The normal cyclic prefix (~4.7 µs) provides a good balance between overhead and supportable RMS delay spread, and is suitable for most delay-spread environments where the RMS delay spread is less than 5 µs which is typical for urban and suburban environments. The extended cyclic prefix is mainly used for very large cells or cells supporting MBMS using a single-frequency network.

The minimum quanta for scheduling downlink data are described by a resource grid of subcarriers and OFDM symbols. The minimum scheduling quanta are defined as a pair of resource blocks (RBs) and described

Table 3.3. *Physical resource-block definition*

Cyclic-prefix type	Number of subcarriers	Number of symbols
Normal	12	7
Extended	12	6

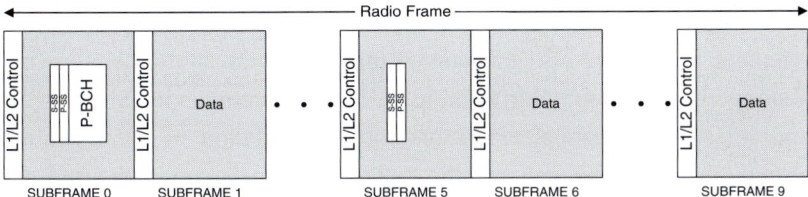

Figure 3.2. FDD downlink subframe structure.

in Table 3.3. The 12 subcarriers span 180 kHz and were chosen to optimize the frequency-selectivity gain. As an example, for normal subframes a resource block consists of 120–144 resource elements (REs). The PDSCH supports QPSK, 16-QAM, and 64-QAM modulation, and is coded using $R = 1/3$ mother turbo code. The turbo code used is the same as Rel-6 UMTS turbo code except that the turbo code's internal interleaver is based on quadratic polynomial permutation (QPP) structure. The shared channel also supports adaptive asynchronous HARQ, adaptive modulation and coding, and MIMO/beamforming with scheduling done at the eNB.

Two frame structure types are supported in LTE – Type 1, which is applicable to FDD deployment, and Type 2, which is applicable to TDD deployment. Type 1 radio-frame structure (FDD) is shown in Figure 3.2. A radio frame is comprised of 10 subframes, each of length 1 ms. This subframe size was chosen to reduce the user-plane latency. The small subframe size allows data to be delivered very quickly and any possible retransmission to be sent soon after the initial transmission. Each downlink subframe is further divided into two 0.5-ms slots to allow interference-randomization techniques (e.g. cell-specific scrambling) to be applied separately to each slot. Interference randomization is a technique in which transmission of the signal from different eNBs is randomized to ensure that users do not always experience interference from a particular eNB. Each slot is comprised of seven OFDM symbols in the case of a normal cyclic prefix or six OFDM symbols in the case of an extended cyclic prefix. Hence there is a loss in throughput due to the lower number of symbols for an extended cyclic prefix. Within each slot, L1/L2 control regions comprising the PCFICH, PHICH, and PDCCH are located in the first one to three OFDM symbols. The PBCH is transmitted in subframe 0 using the central six resource blocks. Similarly, the synchronization signals are transmitted in subframes 0 and 5. In addition, cell-specific reference signals are also transmitted throughout the subframe, allowing a simple channel estimator to be used, like low-complexity MMSE-FIR and IFFT-based channel estimators for demodulation of the downlink shared channel. The cell-specific reference signals are also used for computing channel quality, precoding matrix information, cell search/ acquisition, handover measurements etc.

Type 2 (TDD) frame structure is shown in Figure 3.3. Similarly to Type 1 structure, each radio frame spans 10 ms and consists of 10 1-ms subframes. Since the same spectrum is shared in a TDD deployment, a subframe may be allocated to either downlink or uplink transmission subject to certain constraints. First, subframes 0 and 5 are always downlink subframes, since they contain synchronization signals and broadcast information necessary for users to connect and obtain information about the system. In LTE, two switching-point periodicities are supported – 5 ms and 10 ms. Thus, each TDD uplink/downlink configuration pattern will repeat itself every 5 ms or 10 ms. To provide a switching point between downlink and uplink transmission, a special subframe is used.

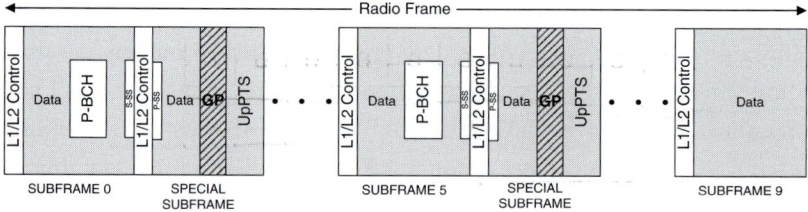

Figure 3.3. TDD frame structure (TDD configuration 1).

This special subframe contains three fields – the Downlink Pilot Time Slot (DwPTS), Guard Period (GP), and Uplink Pilot Time Slot (UpPTS). The DwPTS carries the downlink control and data, similarly to a normal subframe, but with a reduced number of symbols. Thus the DwPTS can be thought of as a shortened downlink subframe. On the other hand, the UpPTS is only provisioned to carry PRACH format 4 (short RACH) and sounding reference symbols. In other words, the UpPTS cannot be used to transmit uplink data and control. The GP in the special subframe provides the total switching period from downlink to uplink transmission and from uplink to downlink transmission. Thus, there is no special subframe to accommodate the switching period from uplink to downlink. Instead, an appropriate timing advance at the UE will be employed to create the necessary guard period for uplink to downlink transition.

For the 5-ms switching-point periodicity, subframes 1 and 6 are special subframes. For the 10-ms switching-point periodicity, however, only subframe 1 is a special subframe, while subframe 6 is a regular downlink subframe. Figure 3.4 illustrates the seven possible DL/UL allocations,

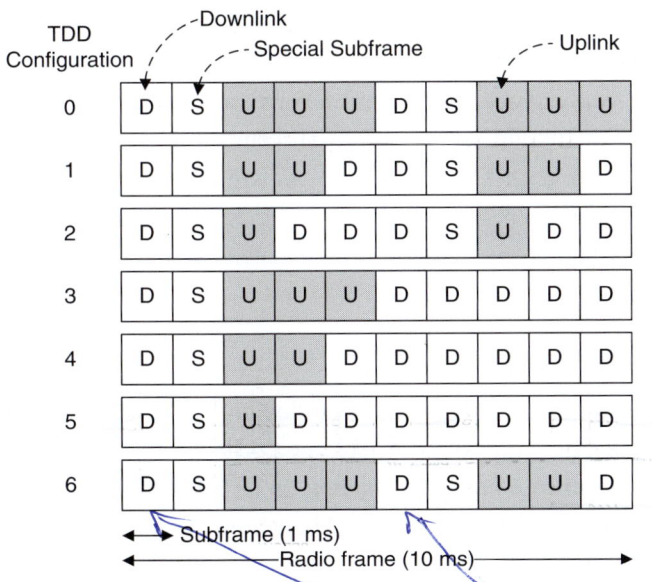

Figure 3.4. TDD DL/UL configurations for LTE.

where D corresponds to a downlink subframe, S corresponds to a special subframe, and U corresponds to an uplink subframe. For instance, DL/UL configuration 1 provides approximately a 60:40 downlink-to-uplink split, while configuration 4 provides approximately an 80:20 downlink-to-uplink time-division split. The choice of DL/UL configurations depends upon the operator's downlink-to-uplink traffic mix and co-existence with other systems such as TD-CDMA or WiMAX. In general, an asymmetric downlink-heavy split is used since many types of data traffic (e.g. web browsing, streaming of video) are by nature directional in favor of the downlink. Note that DL/UL configuration 5, where there is only one uplink subframe, is intended for systems that provide mostly broadcast services.

As illustrated in Figure 3.3, the total length of DwPTS, GP, and UpPTS fields is 1 ms, which is the same as the length of a regular subframe. However, within the special subframe the length of each field may vary depending on the requirement for co-existence with legacy TDD systems, supported cell size, and switching-time requirement. Eight different configurations are defined, providing a DwPTS of varying sizes from 3 to 10 OFDM symbols. For instance, configuration 3 provides a DwPTS:GP: UpPTS of size 11:2:1 OFDM symbols. In this case, 11 OFDM symbols are available for use by the DwPTS, while the UpPTS is of length 1 OFDM symbol. Certain configurations must be used when PRACH format 4 needs to be supported since PRACH format 4 requires UpPTS of size two symbols.

3.4 Data transmission

3.4.1 Shared data channel

 Data transmission on the downlink is transmitted on the PDSCH. Users are allocated resources within the PDSCH via a scheduling assignment contained in the downlink control information. In LTE Rel-8, SU-MIMO and MU-MIMO are both supported. In SU-MIMO, multiple data streams are simultaneously transmitted to the same user using the same physical resource blocks. In MU-MIMO, multiple data streams are simultaneously transmitted to different users using the same physical resource blocks. SU-MIMO requires signaling to inform the users of the relevant transmission parameters, whereas MU-MIMO operation is transparent to the users.

The transmitter chain can be summarized as follows. First, user data is coded using turbo coding with the mother code rate of 1/3 and rate matched to the coding rate assigned by the scheduler. For example, if the user is assigned a coding rate of 1/2, the rate-matching procedure will remove one of the three encoded bits generated using the mother code rate of 1/3 to arrive at the assigned coding rate of 1/2. The encoded bits are next scrambled on the basis of user and cell identities, modulated, precoded, and then mapped to resource elements. The asynchronous HARQ protocol is used in the downlink. This means that retransmission for a specific HARQ process can occur anytime after a predefined delay. For FDD, retransmission for an uplink packet transmitted at subframe n can occur in any subframe after subframe $n + 8$. As a result, control signaling is required for packet retransmission. In addition, packet retransmission may be adaptive or non-adaptive. With non-adaptive retransmission, the UE retransmits the packet using the same resource allocation as assigned previously, with the option of changing only the redundancy version. With adaptive retransmission, the UE may be assigned new downlink resources, modulation and coding rate, and redundancy version, as long as the transport block size is the same as the original.

3.4.1.1 Dynamic and semi-persistent scheduling

Downlink data transmission can be scheduled dynamically every subframe or semi-persistently until the resource is released. For dynamically scheduled data transmission, the scheduling assignment is given via downlink control information (DCI) format 1/1A/1B/2/2A/2B, where the different formats refer to the different types of scheduling assignment. The DCI is transmitted in the downlink control channel, the details of which are provided in the next section.

In LTE, a stop-and-wait ARQ protocol is used to ensure that a data packet is correctly received before a new packet is transmitted. In this protocol, the sender waits for an acknowledgment subsequent to data

ACK /
NACK

transmission. If a positive acknowledgment (ACK) is received, new data is then sent next. However, if a negative acknowledgment (NACK) is received, the previous data packet is retransmitted. To continuously supply the receiver with data, a multi-channel (N) stop-and-wait protocol is used. This protocol parallelizes the stop-and-wait protocol, in effect running N separate instantiations of the HARQ protocol. As a result, the receiver is not stuck waiting for data, since multiple instances of the algorithm send different data blocks on different HARQ process numbers. In LTE the value of N in the case of FDD is set to 8, while for TDD it depends on the TDD DL/UL split. After decoding the downlink data sent in the Nth subframe by the eNB, the UE sends the ACK/NACK at $N+4$ subframes, since the processing time at the UE is 3 ms. On receiving the ACK/NACK, the eNB schedules retransmissions or new transmissions at the subframes with numbers greater than or equal to $N+8$.

Async
HARQ

In the downlink, asynchronous HARQ, in which the UE transmits chase or incremental-redundancy versions of a packet on any HARQ process until an ACK is received or the maximum number of retries has been reached, is used. In the asynchronous HARQ stop-and-wait protocol the UE can send retransmissions on any HARQ process in any subframe, thus allowing greater scheduling flexibility [3]–[4]. For asynchronous HARQ to work, the following needs to be signaled using the DCI format: explicit signaling of the HARQ process, an incremental-redundancy version, and a new data indicator.

Semi-persistent scheduling is suitable for users that have periodic data assignment on the downlink. This mode is designed to save scarce downlink control-channel resources by periodically and persistently allocating downlink resources to users with known periodic downlink data transmission such as VoIP or video conferencing [5]. In this mode a downlink allocation is given and remains available to the user until the allocation is released. As a result, a large number of users can be scheduled in this mode.

3.4.1.2 Resource allocation

Resource blocks are allocated in scheduling assignment grants using an appropriate resource-allocation type (RAT). This assignment informs the UE which resource blocks it has been assigned for data transmission.

Three RATs are available – 0, 1, and 2. All three types can be used for resource assignment in the downlink, but only RAT 2 can be used for uplink assignment. Types 0 and 1 are implemented using bitmaps, where each bit in the map specifies whether the corresponding resource block or resource-block group is assigned to the user. This provides great flexibility as to which resource blocks can be assigned to a given user. However, the disadvantage is that a large number of bits must to be sent to the user, resulting in high control overhead. Figure 3.5 illustrates RAT 0 and 1 for a system bandwidth of 5 MHz. In the case of RAT 0, each bit in the map signifies whether the resource-block group is assigned to the user. The size of the resource-block group is dependent on the system bandwidth. For a system bandwidth of 5 MHz as shown in Figure 3.5, a resource-block group comprises two resources blocks, thus resource can be assigned in multiples of two resource blocks. When RAT 1 is used, an individual resource block can be assigned. Four different RAT 1 patterns can be generated, depending on how the P and shift (S) parameters are set. This allows all different individual resource blocks to be addressed using a reduced number of bits.

RAT 2 provides contiguous virtual or physical allocation of resource blocks. This format requires only the starting resource-block number and the number of assigned resource blocks to be sent to the users. The control overhead with RAT 2 signaling is significantly lower than that for RAT 0/1 signaling. However, scheduling flexibility is lost insofar as only contiguous resource blocks can be assigned to a user. This is the only RAT format that can be used in the uplink, since uplink assignment must be contiguous in order to preserve the single-carrier property.

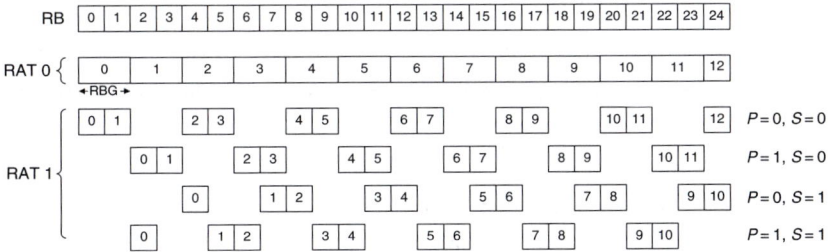

Figure 3.5. Examples of resource-allocation types.

3.4.1.3 Modulation and TBS allocation

Each PDSCH transmission to the UE contains an information set whose size is given by the transport block size (TBS). In order to decode this data packet, the UE needs to know the modulation type, the TBS, and the coding rate. The modulation and TBS are jointly provided to the user in the downlink scheduling assignment using the I_{MCS} field. The coding rate can be determined by the number of resource blocks once the modulation and TBS are known. The possible values are contained in two lookup tables and communicated to the user using the 5-bit MCS Index. The 5-bit index provides 32 possible values, which provide sufficient granularity in modulation and coding-rate assignment while maintaining low overhead. A subset of the table is shown in Table 3.4, and the complete table can be

Table 3.4. *A subset of the MCS index table [6]*

MCS index	Modulation	TBS index
0	QPSK	0
1	QPSK	1
...		
8	QPSK	8
9	QPSK	9
10	16-QAM	9
11	16-QAM	10
...		
15	16-QAM	14
16	16-QAM	15
17	64-QAM	15
18	64-QAM	16
...		
27	64-QAM	25
28	64-QAM	26
29	QPSK	Reserved
30	16-QAM	Reserved
31	64-QAM	Reserved

TBS: Transport Block Size

found in [6]. The table has two entries corresponding to each index value, namely the modulation order of the data packet and the TBS index. As an example, consider an assignment to the user of $I_{MCS} = 10$. From Table 3.4, the user interprets this as stating that 16-QAM modulation will be used to transmit the data, and the TBS of the data packet can be determined from the TBS table using a TBS index of 9.

A subset of the TBS index table is shown in Table 3.5, and the complete table can be found in [6]. Each row in the table provides the size of the transport block in bits for a given TBS index and number of assigned physical resource blocks. The table extends to 110 resource blocks, which is the maximum number of resource blocks supported in LTE. The number of assigned physical resource blocks is determined by the user on the basis of the RAT discussed in Section 3.4.1.2. For instance, with a TBS index of 9, and four assigned resource blocks, the size of the transport block is 616 bits according to Table 3.5. As a result, the user can determine whether it is being scheduled to receive a downlink data packet of size 616 bits (plus 24-bit CRC) that will be transmitted on four resource blocks using 16-QAM modulation. The physical location of those resource blocks is given by a separate resource-allocation field. Using

Table 3.5. *A subset of the TBS index table [6]*

TBS index	Number of physical resource blocks							
	1	2	3	4	5	6	7	8
0	16	32	56	88	120	152	176	208
1	24	56	88	144	176	208	224	256
2	32	72	144	176	208	256	296	328
				. . .				
9	136	296	456	616	776	936	1096	1256
10	144	328	504	680	872	1032	1224	1384
				. . .				
25	616	1256	1864	2536	3112	3752	4392	5160
26	712	1480	2216	2984	3752	4392	5160	5992

this information, the user can also determine the coding rate of the data packet to be sent.

This table is valid for FDD and regular subframes in TDD. For special subframes in TDD, however, the TBS is scaled by 0.75 to account for the reduced number of OFDM symbols. Note that a separate table is used for two-layer spatial multiplexing, whereby two data streams are multiplexed into the same transport block. In that case, the values of the entries are approximately double those in the regular table.

3.4.2 Multimedia broadcast multicast service

Multimedia Broadcast Multicast Service (MBMS) will deliver high-data-rate multimedia services such as audio/video streaming, broadcast data services, and advertising to LTE subscribers. The services are delivered using broadcast or multicast mode. Examples of broadcast services include television and radio broadcasts, service updates, and location-based advertising. Multicast services in general require a sub-scription and can include, for example, personal communications, financial data services, entertainment-related services, and others as offered by the service provider. Because the service is broadcast, the maximum data rate is limited by the signal quality received by the worst user.

In LTE, MBMS is delivered via the PMCH using coordinated multi-cell transmission from a multimedia broadcast single-frequency network (MBSFN) service area. In the MBSFN, the same signal is transmitted from a cluster of neighboring cells belonging to the same MBSFN service area so that the energy in each subcarrier from different cells participating in the MBSFN operation is naturally combined over the air. This is conceptually illustrated in Figure 3.6, where the same multicast channel being transmitted from different cells is coherently combined by the user at the cell edge. This provides a substantial improvement in the SNR at the cell edge, which can increase the maximum supportable data rate substantially. Using the MBSFN, spectral efficiencies at 95% cell cover-age of approximately 1 bps and 3 bps can be achieved for large cells (1732 m site-to-site) and small cells (500 m site-to-site), respectively. In

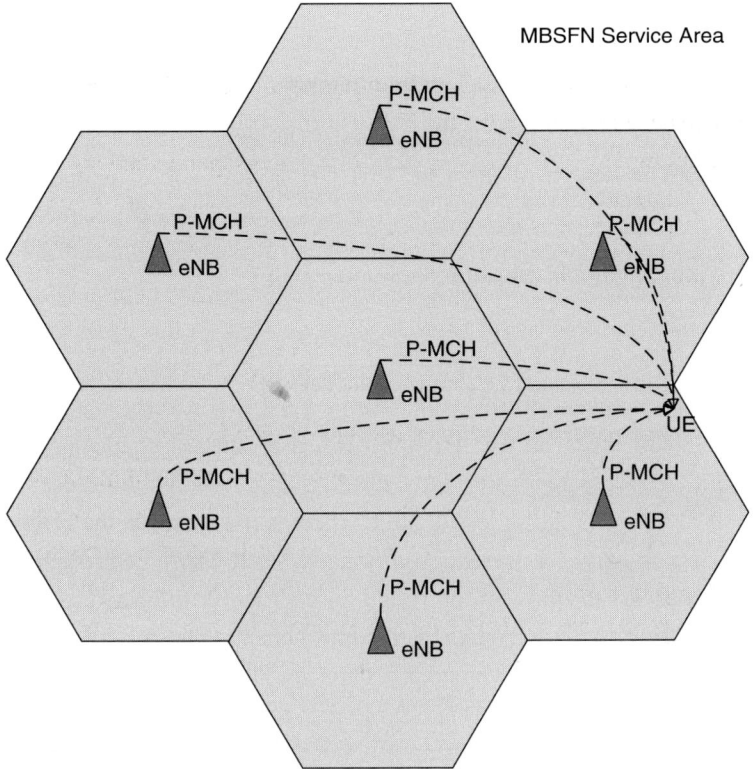

Figure 3.6. MBMS single-frequency network.

this case, data rates of 10–30 Mbps can be supported using a system bandwidth of 10 MHz.

To support MBSFN operation, inter-site synchronization is required. In addition, the cyclic-prefix duration should be long enough compared with the time difference between the signals received from multiple cells. Hence the MBSFN subframe is usually operated using the extended cyclic prefix shown in Table 3.2. The subcarrier spacing of 7.5 kHz using a cyclic-prefix duration of 33 μs is applicable only for standalone MBMS operation using a separate carrier. The MBSFN (PMCH) and unicast traffic (DL-SCH) can also be multiplexed in a TDM fashion with the MBSFN subframe, preferably using an extended cyclic-prefix duration of 16.5 μs.

3.5 Control signaling

3.5.1 Physical Downlink Control Channel

The PDCCH carries the DCI necessary to decode the PDSCH and PUSCH. The design of the LTE PDCCH is very robust and flexible. There are two principal types of control-channel design, namely the broadcast common control channel and the per-user dedicated control channel. In the case of a broadcast control channel, the common control information required by all users is jointly coded. In this case, the code rate used is designed to reach the worst-case user, and the control signals could not be power-controlled or beamformed to a specific user. With a per-user dedicated control channel, each user control channel is independently coded with an individual CRC. Using this structure, a user with a good SINR will require less coding, whereas if a user is at the cell edge the code rate can be lowered. Additionally, each user's control channel can be individually power-controlled as well as beamformed. Analysis shows that with a per-user dedicated control channel the control overhead is reduced and overall coverage and capacity are increased [7]–[8]. This dedicated control structure is similar to that used in Rel-6 HSDPA [9]. In LTE, the downlink control channel is designed on the basis of per-user dedicated information for which the power and coding rate can be individually controlled. Multiple control channels are used, and a user monitors a number of control channels in order to find its dedicated control information. Each channel carries information associated with a radio-network temporary ID (RNTI), which is the identity assigned to a UE by the network. For the control channel, convolutional code is used due to its good performance for the small packet size typical of the control information. Different code rates are supported through the use of different numbers of control-channel resource elements.

The PDCCH is located in the first n OFDM symbols of the subframe, where n is dependent on the bandwidth, subframe, and service type [10]. In a typical FDD deployment, n is less than or equal to 3, and is dynamically signaled to the users via the PCFICH. Data transmission begins after n OFDM symbols, thus there is no mixing between data and control. Placing the control information in the first n symbols of the subframe means that micro-sleep power-saving mode can be employed by the users. In this mode, the receiver can wake up during transmission of the control information and, if there is no assignment, go back to sleep within one symbol, giving battery-life savings of approximately 64%.

There are 10 available DCI formats in LTE Rel-8. Within each DCI format, several types of control information are sent – information related to resource indication such as the resource-block size and duration of assignment, information related to the transport format such as multi-antenna information, modulation scheme, and payload size, and information related to HARQ support such as the process number, the redundancy version, and a new data indicator. The available formats are briefly described in Table 3.6.

The detailed bit allocation for some of the DCI formats used for downlink scheduling assignment is shown in Tables 3.7–3.9. Details for uplink scheduling assignment (DCI format 0) are discussed in Section 4.3.1.

Table 3.6. *DCI formats*

Format	Description
0	Uplink scheduling assignment
1	Downlink scheduling assignment for SIMO
1A	Compact downlink scheduling assignment for SIMO
1B	Based on format 1A with the addition of 2–4 bits for PMI, 2 bits for two TX antennas and 4 bits for four TX antennas
1C	Very compact downlink assignment for one codeword
1D	Compact downlink assignment for scheduling of one PDSCH codeword with precoding and power-offset information
2	Downlink scheduling assignment for SU-MIMO
2A	Downlink scheduling assignment for open-loop spatial multiplexing
3	Transmission of TPC commands for PUCCH and PUSCH with 2-bit power adjustments
3A	Transmission of TPC commands for PUCCH and PUSCH with 1-bit power adjustments

Table 3.7. *DCI format 1, downlink scheduling assignment for SIMO*

Field	Number of bits	Description
Resource allocation header	1	
RB assignment	$\lceil N_{RB}^{DL}/P \rceil$	DL resource-block assignment
MCS	5	Modulation and coding level
HARQ process number	3 (FDD), 4 (TDD)	HARQ ID for asynchronous N-channel stop-and-wait
New data indicator	1	
Redundancy version	2	HARQ based on incremental redundancy
TPC for PUCCH	2	
DL assignment index	2	Applies to TDD only
CRC + UE ID	16	

The purpose of the multiple DCI formats is to scale the number of bits for the PDCCH depending upon the UE and eNB features. As an example, if the downlink spatial multiplexing feature is not invoked in the eNB there is no need for scheduling with the control-channel format for spatial multiplexing (format 2), since this requires a greater number of control-channel bits. There are four types of downlink assignment corresponding to one codeword and two types of downlink assignment corresponding to two codewords. There are two types of assignment – localized and distributed. Distributed transmission is generally associated with frequency-diverse scheduling, whereas localized transmission is associated with frequency-selective or frequency-non-selective scheduling.

The new data indicator (explicitly based on the NDI field) clears the HARQ buffer and starts decoding the data according to the contents of assignments when detected by the UE. After receiving a downlink scheduling assignment, the UE sets up the HARQ buffer on the basis of the contents of the MCS and the resource-block assignment field. Since the downlink HARQ transmission is based on an asynchronous

Table 3.8. *DCI format 1A, compact downlink scheduling assignment for SIMO*

Field	Number of bits	Description
Format differential flag	1	
Localized/ distributed VRB assignment flag	1	Localized or distributed resource-block allocation
RB assignment	$\lceil \log_2(N_{RB}^{DL}(N_{RB}^{DL}+1)/2) \rceil$	
MCS	5	Modulation and coding level
HARQ process number	3 (FDD), 4 (TDD)	HARQ ID for asynchronous N-channel stop-and-wait
New data indicator	1	
Redundancy version	2	HARQ based on incremental redundancy
TPC for PUCCH	2	
DL assignment index	2	Applies to TDD only
CRC + UE ID	16	

stop-and-wait protocol, the HARQ process number field is needed for the downlink, whereas it is not required for the uplink because of the use of the synchronous stop-and-wait HARQ.

Each scheduling grant is defined on the basis of fixed-size control-channel elements (CCEs), where a CCE is defined as a group of resource elements. The CCEs can be combined in a predetermined manner to achieve different coding rates. A CCE aggregation level of one, two, four, or eight CCEs can be used to support one DCI transmission. Each CCE is made of nine resource element groups, where each group contains four resource elements. The resource elements used within the CCE are distributed throughout the entire control-channel region, providing

Table 3.9. *DCI format 2, downlink scheduling assignment for MIMO*

Field	Number of bits	Description
Resource-allocation header	1	
RB assignment	$\lceil N_{RB}^{DL}/P \rceil$	DL resource-block assignment
TPC for PUCCH	2	
DL assignment index	2	Applies to TDD only
HARQ process number	3 (FDD), 4 (TDD)	HARQ based on incremental redundancy
Transport block to codeword swap flag	1	
Transport block 1		
MCS	5	Modulation and coding level
New data indicator	1	
Redundancy version	2	HARQ based on incremental redundancy
Transport block 2		
MCS	5	Modulation and coding level
New data indicator	1	
Redundancy version	2	HARQ based on incremental redundancy
Precoding information	3 or 6	
CRC + UE ID	16	

frequency-diversity gain and an interference-averaging effect. QPSK modulation is always used. Because multiple CCEs can be combined to effectively reduce the effective coding rate, a user control-channel assignment would then be based on channel-quality information reported. For instance, with DCI format 1 the effective coding rate is 0.53 with one CCE and 0.07 with eight CCEs.

The UE uses blind decoding to detect the PDCCH. To reduce the number of blind detections, two search spaces are defined – the common search space and the UE-specific search space. In the UE-specific search space, the UE searches aggregation levels 1, 2, 4, and 8, whereas in the common search space the UE searches aggregation levels 4 and 8. The common search space begins at CCE index 0. The UE determines the starting CCE of the UE-specific search space using a hashing function that is dependent on the MAC identity and subframe number.

3.5.2 Physical Control Format Indicator Channel

The PCFICH is used to dynamically indicate the number of OFDM symbols used for control in a subframe. The possible values are dependent on the subframe type, structure, and system bandwidth. For system bandwidth greater than 1.4 MHz, up to three OFDM symbols may be reserved for control signaling. Further restriction is placed, however, on MBSFN subframes to limit the maximum number of control symbols to two. This restriction is enforced also for subframes 1 and 6 of the frame structure of Type 2 (TDD) since the primary synchronization signal is transmitted on the third OFDM symbol. For system bandwidth of 1.4 MHz, up to four symbols may be used for control signaling since the bandwidth of the control region is very small. This will ensure that one has a sufficiently large control region to schedule multiple uplink and downlink users in the same subframe.

The PCFICH is transmitted in the first OFDM symbol of the subframe and has three possible values. Because of the small number of information bits carried by the control format indicator, possible coding options are limited. As a result, predefined codewords based on (3,2) simplex coding with repetition and systematic bits with minimum Hamming distance of 21 are used, as shown in Table 3.10. The selected codeword is mapped to 16 QPSK symbols, which are subsequently mapped to different resource elements. To provide maximum frequency diversity, the PCFICH is mapped onto four resource-element groups that are uniformly distributed over the whole system bandwidth. The mapping ensures that each group is separated by approximately a quarter of the system bandwidth, ensuring that the channel seen by each resource-element group is as uncorrelated as possible. Note that the PCFICH mapping to resource elements is fixed and known from the cell identity. Transmit diversity is also supported using the same diversity scheme as the PDCCH. This combination

Table 3.10. *Control format indicator codeword*

Value	Codeword
1	0, 1, 1, 0, 1, 1, 0, 1, 1, 0, 1, 1, 0, 1, 1, 0, 1, 1, 0, 1, 1, 0, 1, 1, 0, 1, 1, 0, 1, 1, 0, 1
2	1, 0, 1, 1, 0, 1, 1, 0, 1, 1, 0, 1, 1, 0, 1, 1, 0, 1, 1, 0, 1, 1, 0, 1, 1, 0, 1, 1, 0, 1, 1, 0
3	1, 1, 0, 1, 1, 0, 1, 1, 0, 1, 1, 0, 1, 1, 0, 1, 1, 0, 1, 1, 0, 1, 1, 0, 1, 1, 0, 1, 1, 0, 1, 1

of frequency and transmit diversity allows the PCFICH to have good performance and robustness against noise and interference. In a typical urban propagation channel, the SNR requirement for PCFICH performance at 0.1% error rate is around $-2\,$dB. If the operating SNR at the cell edge is below this threshold, power boosting (i.e. transmitting the subcarriers used for PCFICH at higher power than other subcarriers) can be used to ensure good PCFICH performance. Because only 16 resource elements are used for the PCFICH, power boosting of this channel can easily be done if needed.

3.5.3 Physical HARQ Indicator Channel

The PHICH carries the acknowledgment response to uplink transmission on the PUSCH by the UE. Subsequent to transmission of an uplink transport block on the PUSCH, the UE will receive an acknowledgment of the packet on a specific PHICH resource on the basis of a predefined timing. The acknowledgment may be positive (ACK), meaning that the data transport block was received and decoded correctly, or negative (NACK), meaning that the packet was not decoded correctly. Upon NACK reception, the UE can retransmit the data packet at a predefined time. In the physical layer, the acknowledgment is transmitted on a PHICH resource, which is defined by an index pair comprising the PHICH group number $n_{\mathrm{PHICH}}^{\mathrm{group}}$ and the sequence number within the group $n_{\text{PHICH}}^{\text{seq}}$. Each PHICH group contains eight different sequences, in the case of a normal cyclic prefix, or four different sequences, in the case of an extended cyclic prefix.

The UE determines the PHICH resource associated with its uplink transmission by the lowest resource-block index for uplink allocation and the 3-bit cyclic-shift parameter for the uplink reference signal given in the grant. By using the lowest resource-block index for uplink allocation, the PHICH resource can be implicitly determined. This provides a saving insofar as explicit PHICH resource assignment is not required. However, to be able to handle spatial multiplexing where multiple users can share the same lowest resource block, an explicit parameter, namely the 3-bit cyclic-shift parameter for the DMRS, is also used. This ensures that the PHICH resource assignment is free from contention. For instance, for a user given an uplink assignment of four resource blocks PRB = 5, 6, 7, 8 and $n_{\text{DMRS}} = 1$, the corresponding PHICH resource is indicated by $n_{\text{PHICH}}^{\text{group}} = 6$ and $n_{\text{PHICH}}^{\text{seq}} = 2$. If another user is scheduled on the same set of resource blocks but with $n_{\text{DMRS}} = 4$, then that user's PHICH resource is indicated by $n_{\text{PHICH}}^{\text{group}} = 2$ and $n_{\text{PHICH}}^{\text{seq}} = 4$.

To allow control of the PHICH overhead, the number of PHICH groups can be configured by the *phich-Resource* RRC parameter, which can take on values of {1/6, 1/2, 1, 2}. The possible values specify the number of PHICH resources as a multiple of the number of resource blocks. For instance, a value of 1/2 means that approximately half the PHICH resource exists as physical resource blocks, whereas a value of 2 means that there is approximately twice as much PHICH resource as physical resource blocks. By restricting the number of PHICH groups to a minimum, more resource elements are made available for the PDCCH to transmit downlink control information. When there is less PHICH resource than the number of physical resource blocks, the eNB is responsible for ensuring that there is no PHICH-resource contention among the users. The eNB does this by checking for PHICH-resource contention when scheduling a user. If no PHICH-resource is available, the eNB may be prevented from scheduling a particular user. An example of the number of PHICH resources for a subframe with a normal cyclic prefix is shown in Table 3.11. From this table, it can be seen that, with *phich-Resource* set

Table 3.11. *Number of PHICH resources (normal cyclic prefix)*

System bandwidth (MHz)	1/6	1/2	1	2
5	8	16	32	56
10	16	32	56	104
20	24	56	104	200

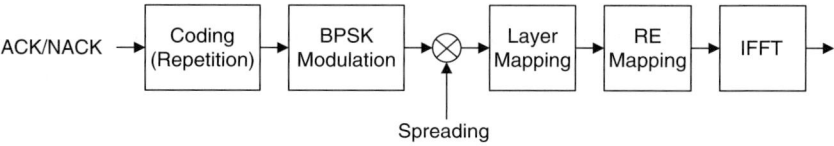

Figure 3.7. PHICH block diagram.

to 1/2, 56 unique PHICH-resource indices are available at 20 MHz. This means that up to 56 uplink users can be simultaneously scheduled in the same subframe.

An illustrative block diagram of the PHICH is shown in Figure 3.7. Since Rel-8 of the specifications does not support spatial multiplexing from the same user on the uplink, only one acknowledgment bit is to be transmitted. The acknowledgment bit is first coded using repetition to form a block of bits. These coded bits are then modulated using BPSK modulation and spread by an orthogonal sequence. For a normal cyclic prefix, eight orthogonal sequences are available using a spreading factor of 4 together with I/Q multiplexing. For an extended cyclic prefix, four orthogonal sequences are available using a spreading factor of 2 together with I/Q multiplexing. This allows multiple acknowledgments to be code-multiplexed into the same PHICH group. Code-division multiplexing increases the PHICH capacity, allows power control between acknowledgments for different users, and provides good interference averaging. Subsequent to the spreading operation, layer mapping using transmit diversity is performed on the basis of the number of transmit antennas at the eNB.

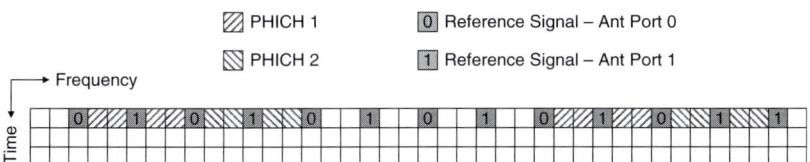

Figure 3.8. PHICH mapping (normal duration).

The mapping of PHICH in the time–frequency domain is illustrated in Figure 3.8. Each PHICH group is comprised of three repeated segments that are mapped to different frequency regions to provide frequency diversity. In addition, the PHICH group can be mapped into one (normal duration) or three (extended duration) OFDM symbols. Mapping into three OFDM symbols allow more power to be allocated to the PHICH, increasing the coverage and reliability of this channel.

For the PHICH, the performance requirement is usually specified as an ACK-to-NACK error probability of 0.1%. Because each PHICH group contains multiple acknowledgments, the power requirement depends on the number of acknowledgments. When only a single acknowledgment is present in the PHICH group, the SNR requirement is around -2 dB in a typical urban channel. This makes the PHICH a very robust channel insofar as it can operate well at the SINR typically seen at the cell edge. With multiple acknowledgments in a PHICH group, interference among different acknowledgments can degrade performance somewhat. However, the eNB can use intelligent scheduling management (e.g. by limiting the number of acknowledgments within a PHICH group or ensuring that all acknowledgments within a group are transmitted with similar powers) to ensure minimal interference.

3.5.4 Physical Broadcast Channel

In LTE, system information necessary for system access is divided into two parts – the *MasterInformationBlock* (MIB) and a number of *SystemInformationBlock* (SIBs). The MIB contains basic physical-layer system parameters necessary to demodulate downlink channels. Subsequent to successful acquisition of the MIB information, the UE

Table 3.12. *Master-information-block fields*

MIB field	Number of information bits
Downlink system bandwidth	3
PHICH configuration	3
System frame number	8
Spare	10

will be able to read the downlink control and data channels, which would allow it to obtain additional control information contained in the SIBs that is necessary in order to perform initial access into the system. These SIBs are transmitted on the data channel. The MIB is transmitted on the PBCH over four radio frames (40 ms period) with portions transmitted in subframe 0 of every radio frame. The MIB contains four fields as shown in Table 3.12. The downlink system bandwidth gives the number of resource blocks in the downlink with possible values of 6, 15, 25, 50, 75, and 100 (corresponding to system bandwidths of 1.4, 3, 5, 10, 15, and 20 MHz, respectively). The PHICH configuration gives the amount of PHICH resource and its duration. The system frame number (SFN) field provides the most significant 8 bits of the 10-bit SFN. The last two bits are to be determined blindly by users on the basis of the coding structure of the BCH.

A block diagram of the BCH transmission is shown in Figure 3.9. First, a 16-bit CRC is appended to the BCH transport channel, forming an information block of size 40 bits. Note that the CRC for this transport is further masked by specific codewords denoting the number of transmit antennas (1, 2, or 4) at the eNB. Thus, by decoding the BCH, the UE will receive confirmation regarding the number of eNB transmit antennas. Prior to this confirmation, the UE will need to blindly determine the number of transmit antennas via the synchronization signals. The transport block is then coded using $R = 1/3$ tail-biting convolution coding. Rate-matching is next performed to generate 1920 coded bits for a normal cyclic prefix and 1728 coded bits for an extended cyclic prefix. This provides a very low coding rate (e.g. a coding rate of 1/48 for a normal cyclic prefix if all transmissions are considered), and as a result the BCH

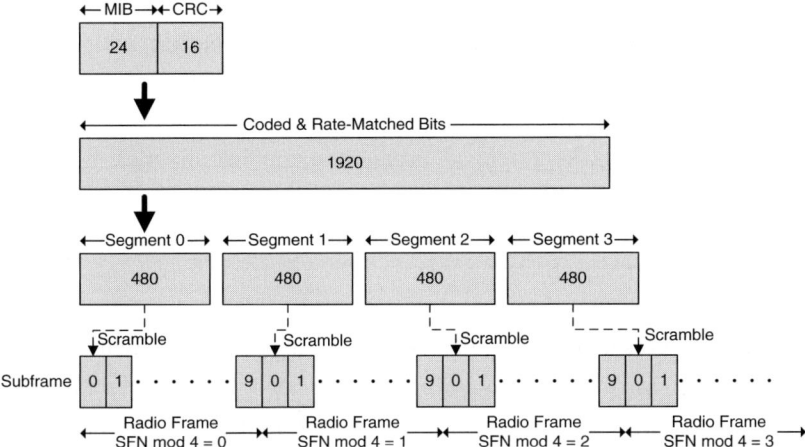

Figure 3.9. Mapping of BCH into four 10-ms subframes (normal cyclic prefix).

enjoys robust performance even for low SNRs. The coded bits are then scrambled and equally divided into four blocks and mapped into four consecutive radio frames. This is illustrated in Figure 3.9. The scrambled sequence is initialized by the cell ID every four radio frames, and therefore is different for each of the segments shown. The position of the PBCH is always fixed in time to the first four OFDM symbols of slot 1, subframe 0 of every radio frame, and in frequency to 72 subcarriers centered around the DC subcarrier. QPSK modulation is always used. When more than one transmit antenna is present at the eNB, transmit diversity is used for antenna diversity. In FDD, the PBCH is placed next to the synchronization signals in the transmission subframe. This is because the UE will generally use channel estimation obtained from the synchronization signals to demodulate the PBCH since the downlink bandwidth is not known. Therefore, by having the PBCH next to the synchronization signals in time, good channel estimates can be ensured. In TDD, the PBCH is placed slightly ahead in time of the synchronization signals. This allows the user to be able to distinguish between FDD and TDD systems during the cell-search stage.

Each of the four PBCH coded segments is self-decodable. However, since each segment has a different scrambling sequence, the UE will have to try four decoding possibilities at each instance. This, however, allows the UE to implicitly derive the last two bits of the SFN from which of the hypotheses was successful. Note that, if the UE cannot decode the PBCH in one reception, it can combine several subsequent transmissions to increase the SNR. In a typical urban propagation channel, the SNR requirement for PBCH performance of 1% BLER is around -8 dB. This makes the PBCH a very robust channel.

3.5.5 Paging Control Channel

The PCH is used either to transmit paging information to UEs in idle mode or to inform both idle and connected UEs about a system information change or Earthquake and Tsunami Warning System (ETWS) notification. UEs that are in idle mode may be configured with a specific discontinuous reception (DRX) cycle, and thus will wake up only intermittently to monitor the Paging Control Channel (PCH). DRX allows the UE to turn off its electronics while in the discontinuous reception state, providing battery-life saving. The paging message contains the following fields – *systemInfoModification* field, to indicate a change in system information, *etws-Indication* field, to indicate ETWS notification, and a list of paging records. Each paging record contains the UE identity and the originating domain for paging. Up to 16 paging records can be grouped together in one paging message, thus 16 UEs can be addressed at the same time.

The paging message is transmitted on the Paging Control Channel. This is a logical channel that is transmitted onto the PDSCH using a special identifier called the Paging Radio Network Temporary Identifier (P-RNTI). Only one P-RNTI value is available and this is shared by all users. In this case, the eNB transmits the paging message on the PDSCH together with an associated downlink scheduling assignment using either DCI format 1A or DCI format 1C. Because this is transmitted on the PDSCH, the eNB has the flexibility to select the amount of resources (modulation and coding, and number of resource blocks) sufficient to reach the intended UEs. Because this message is normally addressed to multiple UEs, ACK/NACK is not possible and therefore the eNB will generally transmit this message with some performance margin to ensure reception. UEs that have been configured for paging notification (i.e. users that are in idle mode) will decode the control information and then the paging message. If the *systemInfoModification* or *etws-Indication* is set, the UE will reacquire the system information and perform the appropriate action. In the case of a match of its identity in the paging record, it will forward this information to the upper layer in preparation for the establishment of an RRC connection to receive an incoming call.

3.6 Downlink reference signal

Downlink reference signals are provided to assist the users in performing channel estimation and demodulation of control and data information, and for channel information measurements such as channel quality information (CQI), rank, and precoding matrix indicator (PMI). Three types of downlink reference signals are available – cell-specific reference signals, MBSFN reference signals, and dedicated reference signals. Cell-specific reference signals are common to all users, while UE-specific reference signals are used only with dedicated reference-symbol-based beamforming. MBSFN reference signals are used in MBSFN transmission and are common to all cells in the MBSFN service area. In LTE, a reference-signal pattern is specified per downlink antenna port, where an antenna port is a logical entity that can be mapped to one or more physical antennas. For cell-specific and MBSFN reference signals, one, two or four antenna ports are available. As an example, an eNB with two physical transmission antennas can be configured to use either one or two antenna ports. If it is configured to support two antenna ports, different reference signal patterns are transmitted on the two available physical antennas. If only one antenna port is used, then the same reference-signal pattern is transmitted on both physical antennas. The number of cell-specific antenna ports must be first determined blindly by the UEs, and then confirmed via the *antennaPortsCount* field of the radio resource configuration message. For UE-specific reference signals, however, only one antenna port (antenna port 5) is defined. In this case, only one reference-signal pattern is defined, which is then mapped to all available physical antennas.

The downlink cell-specific reference-signal structure is shown in Figure 3.10 for four antenna ports. Note that, for a smaller number of antenna ports, the unused reference-signal symbols are available for either data or control transmission. For instance, with one antenna port, resource elements for reference signals R_1, R_2, and R_3 are used for either data or control transmission. The overhead consumed by the reference signals is 4.8%, 9.5%, and 14.3% for one, two, and four antenna ports, respectively. To provide good channel-estimation performance, cell-specific reference symbols are provided at regular time and frequency intervals. From Figure 3.10, it can be seen that there is a frequency separation between reference symbols of six resource elements. This gives a coherence bandwidth of 90 kHz, which can support a propagation channel with RMS delay spread of up to 2.2 μs. In the case of one or two antenna ports, reference symbols are present in four OFDM symbols that are distributed fairly evenly within the resource block. This temporal separation provides a coherence time of approximately 0.3 ms and can

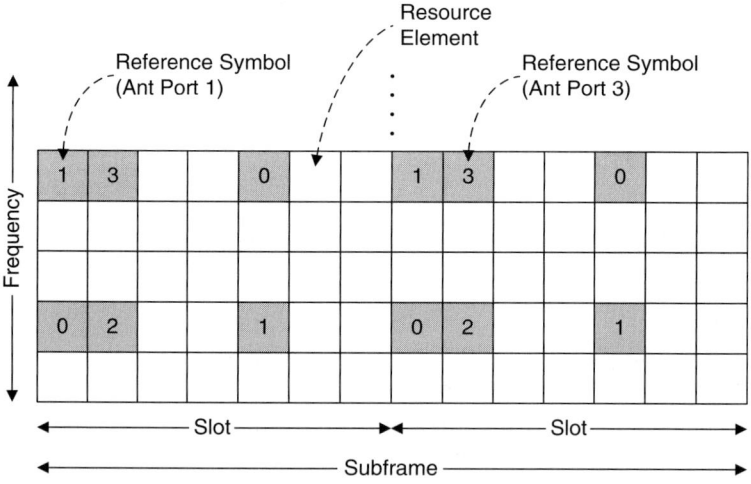

Figure 3.10. Mapping of the cell-specific common downlink reference signals (normal cyclic prefix).

therefore easily support channel estimation for UE with speed up to 350 km/h. In the case of four antenna ports, however, reference symbols are present in only two OFDM symbols within the resource block in order to reduce the overhead. In this case, channel estimation is good for UE speed up to 120 km/h but merely functional at 350 km/h.

The cell-specific reference signals are generated using a pseudo-random sequence and mapped on all reference-symbol positions as shown in Figure 3.10. The pseudo-random sequence is initialized with the cell identity, slot and symbol number, and cyclic-prefix type. Thus, the reference signal will be different in each OFDM symbol, providing a measure of interference randomization across different cells. In addition, the frequency-domain position of the cell-specific reference signals is determined by the cell identity. Since the frequency separation between reference-signal positions is six resource elements, up to six different shifts are available. This allows, for example, support for six sectors or cells per eNB, each with a unique frequency positioning of the reference signals to minimize interference.

The reference-signal structure for MBSFN subframe using an extended cyclic prefix is shown in Figure 3.11. Only one antenna port (antenna port 4) is used, which means that only one reference-signal pattern is defined and all physical transmit antennas will use this pattern. From

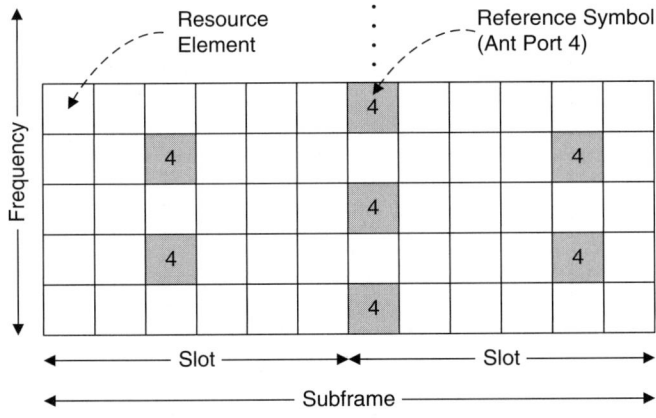

Figure 3.11. Reference-signal structure for MBSFN subframes.

Figure 3.11, it can be seen that the reference symbol density in both time and frequency is much larger than that for the cell-specific reference signals. This is because much finer frequency resolution is needed in the case of MBSFN to provide good channel estimation due to the highly frequency-selective nature of the MBSFN signals which are transmitted from different cells within the MBSFN service area. High frequency-selectivity means that the channel changes quickly in the frequency domain. Thus, finer frequency resolution of the reference signal is needed in order to provide good channel estimation. The reference-signal over-head in this case is 12.5%. In the MBSFN subframe, up to two of the first OFDM symbols can be reserved for non-MBSFN transmission and will not be used for PMCH transmission. The first two symbols can be used for uplink assignment and power control, measurement purposes such as CQI, and channel tracking. It may be noted that PMCH is not transmitted in subframes 0 and 5 on a carrier supporting a mix of PDSCH and PMCH transmission. Similarly to the cell-specific reference-signal sequence, the MBSFN reference-signal sequence is generated using a pseudo-random sequence that is initialized by the OFDM symbol number and MBSFN identity. There is, however, no cell-specific shift because signals from all eNBs in the MBSFN service area must be coherently combined.

Finally, the UE-specific reference-signal structure for LTE Rel-8 is shown in Figure 3.12. The user-specific reference signals are used for UE-specific beamforming and consist of 12 dedicated symbols per resource block (7.1% overhead). In addition, cell-specific reference signals are also needed in order for the users to demodulate control information. For instance, with two cell-specific antenna ports and UE-specific reference signals, the overall overhead is 16.6%. As a result, the number of antenna ports used for cell-specific reference signals should be reduced to one or two in order to keep the overhead reasonable. In that case, control informa-tion can be transmitted using either transparent or standards-based transmit-diversity schemes. If a UE is transmitting in beamforming mode, both the PDSCH and the dedicated reference signals are weighted by the same beamforming weights computed at the eNB. Although only one pattern is defined, different beamforming weights are applied to each antenna and thus different signals are transmitted out over the physical antennas.

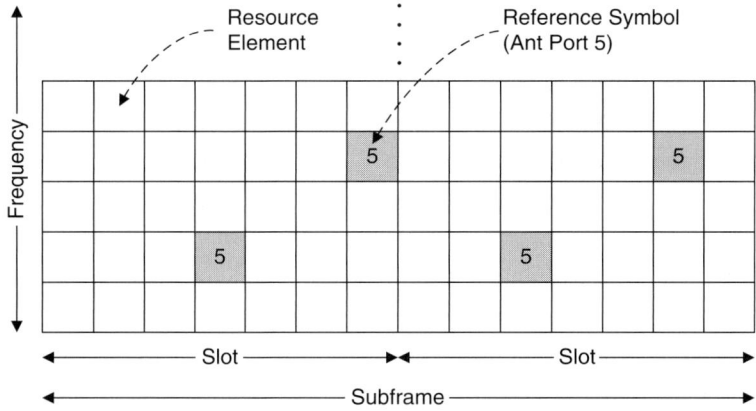

Figure 3.12. UE-specific reference-signal structure (normal cyclic prefix).

3.7 Synchronization signals

Downlink synchronization signals are used for a variety of purposes, including detection of frame timing, initial and neighbor-cell acquisition, computing the number of antennas in the BCH, and for handover purposes. It may also be noted that the neighbor-cell search is based on the same downlink signals as for the initial cell search. Two synchronization signals are available in LTE – the primary synchronization signal (PSS) and the secondary synchronization signals (SSS). Both the PSS and the SSS are transmitted on 72 active subcarriers located around the DC subcarrier. For FDD, the PSS and SSS are transmitted on subframes 0 and 5 and occupy two symbols in a subframe as shown in Figure 3.2. For the TDD frame structure, the SSS is transmitted on the last symbol of subframe 0 as illustrated in Figure 3.3. The PSS is transmitted on the third OFDM symbol in the DwPTS to allow the same reference-signal placement in the DwPTS as in other downlink subframes. This, however, will result in a small degradation of the cell-search performance using coherence detection at very high vehicular speed due to channel variations, since the PSS and SSS are not located next to each other.

The synchronization signals are tied to the physical-layer cell identity. Users first obtain the group identity $N_{ID}^{(1)} = (0, \ldots, 167)$ from the sequence index used for the PSS. Next, the identity within the group

$N_{\text{ID}}^{(2)} = (0, 1, 2)$ is obtained from the sequence index used for the SSS. Upon detection of the synchronization signals, users can determine the physical-layer identity (also known as the cell identity) using the relationship $N_{\text{ID}}^{\text{cell}} = 3N_{\text{ID}}^{(1)} + N_{\text{ID}}^{(2)}$. Since there are 168 unique groups and 3 unique identities within each group, 504 unique physical-layer identities are available [11]. This provides approximately the same number of physical-layer identities as in UMTS systems. Through careful cell planning of physical-layer cell identity, it is possible to optimize initial and neighbor-cell search performance by ensuring that neighboring cells employ synchronization signals with minimum correlation. A common approach is for the sectors within a cell to share the same group identity but with different cell identities.

3.7.1 Cell search and synchronization sequences

Cell search in LTE is performed in a hierarchical procedure involving detection of the two synchronization signals [12]. This two-step, hierarchical cell-search procedure was chosen due to its robust performance in low-SNR environments. In the first step, acquisition of the symbol timing necessary to align frequency processing is done on the basis of the PSS. This is generally done through correlating the received signal with a known set of PSSs. Subsequent to symbol-timing acquisition, radio frame timing and cell identity are acquired from the SSSs. This is usually done using more efficient frequency-domain correlation techniques.

The sequence used for the PSS is generated using a Zadoff–Chu sequence with a different root sequence index. Zadoff–Chu sequences have good auto-correlation and very low cross-correlation, thus minimizing interference. In addition, a Zadoff–Chu sequence has a constant amplitude, which results in a low peak-to-average power ratio in the transmitter, has low implementation complexity, and can provide a phase reference that can be used to aid in the detection of the SSS. Although the 72 central subcarriers are used for the synchronization signal, the sequence is of length 63 since the length of the Zadoff–Chu sequence should be a prime number for optimal performance. The remaining nine subcarriers are nulled (i.e. there is no transmission on those subcarriers). The parameter $N_{\mathrm{ID}}^{(2)} = (0, 1, 2)$ corresponds to using an index of 25, 29, and 34, respectively. These indices were chosen because they allow all three possible PSSs to be quickly processed together and also because these sequences result in superior performance. The SSS sequence is generated by concatenating two length-31 binary sequences. The combination of these two sequences differs between subframe 0 and subframe 5, which helps to detect radio frame timing and cell identity.

3.8 Performance results

3.8.1 Link-level performance

Table 3.13 provides an illustrative example of the typical operating requirements and associated SNR for various downlink channels. For example, for the PHICH, a typical operating point is for the false-alarm

Table 3.13. *Downlink link-level channel performance (two transmit and two receive antennas)*

Physical channel	Typical operating requirements	Typical SNR operating points (dB)
PCFICH	1% error	−2.1
PHICH	P(miss) < 0.1%	−2.0
PBCH	1% BLER	−8.0
PDCCH		
DCI format 0, eight CCEs	1% BLER	−5.4
DCI format 2, two CCEs	1% BLER	−3.3
PDSCH		
VoIP, 12.2 AMR	10% BLER for first transmission	−5.5
FTP, 5.2 Mbps	10% BLER for first transmission	1.0
FTP, 21.4 Mbps	10% BLER for first transmission	10.0
FTP, 36.7 Mbps	10% BLER for first transmission	17.5

and false-detection rate to be below 0.1%. For the data channel, the operating point is usually set at 10% block error rate (BLER) for the initial transmission, resulting in throughput of approximately 91% of the initial data rate when HARQ is considered. In Table 3.13, the required SNR per subcarrier is given for the typical urban (TU) propagation channel [3], which is widely used in analysis of cellular networks due to its accurate representation of an urban propagation environment. At the eNB, two transmit antennas are used with transmit diversity. Two receive antennas are present at the UE, providing receiver diversity as well as combining gains. Additional link-level performance results may be found in [13].

3.8.2 System-level performance

In this section, the downlink system performance of LTE Rel-8 is provided as per the ITU scenarios using transmit diversity, with two transmit antennas and two receive antennas at the eNB and UE, respectively, and using a full-buffer-traffic model. It may be noted that the performance of the downlink with various multi-antenna schemes is shown in Chapter 5. The downlink system performance is provided for four distinct scenarios defined in [14] – an indoor hotspot, an urban micro-cell, an urban macro-cell, and a rural macro-cell. The baseline configuration parameters are shown in Table 3.14.

Non-ideal channel estimation is assumed and the subband CQI is reported every 5 ms. Localized allocation (using frequency-selective scheduling) is simulated for 2×2 SFBC. The system performance for 10 UEs per sector for full buffer traffic using frequency-selective scheduling is summarized in Table 3.15 [15]. Full buffer traffic guarantees that there will always be data available for each user. The corresponding simulation assumptions are summarized in Table 8.1.

It may observed from Table 3.15 that, with 2×2 operation, the E-UTRA downlink sector spectral efficiency is highest for an indoor environment, while for the other environments it varies from 1.3 to 1.7 bps/Hz. This shows that the transmit diversity is quite robust with respect to vehicle speed and different environments. Additional downlink system performance results may be found in [16]–[19].

Table 3.14. *Simulation scenarios*

Deployment scenario for the evaluation process	Urban macro-cell	Urban micro-cell	Indoor hotspot	Rural macro-cell
Total eNB transmit power	46 dBm for 10 MHz, 49 dBm for 20 MHz	41 dBm for 10 MHz, 44 dBm for 20 MHz	24 dBm for 40 MHz, 21 dBm for 20 MHz	46 dBm for 10 MHz, 49 dBm for 20 MHz
Inter-site distance (m)	500	200	60	1732
User distribution, randomly and uniformly distributed over area	100% of users outdoors in vehicles	50% of users outdoors (pedestrian users) and 50% of users indoors	–	100% of users outdoors in high-speed vehicles
UT speeds of interest (km/h)	30	3	3	120
BS noise figure (dB)	5	5	5	5
UT noise figure (dB)	7	7	7	7
Thermal-noise level (dBm/Hz)	−174	−174	−174	−174
Carrier frequency (GHz)	2	2.5	3.4	0.8
Number of eNB antennas	Up to eight transmitters/receivers			
Number of UE antennas	Up to two transmitters/receivers			

Table 3.14. (*cont.*)

Deployment scenario for the evaluation process	Urban macro-cell	Urban micro-cell	Indoor hotspot	Rural macro-cell
UE power class (dBm)	24	24	21	24
Outdoor to in-car penetration loss (dB)	9 (LN, $\sigma = 5$ dB)	–	–	9 (LN, $\sigma = 5$ dB)

Table 3.15. *System performance using 2 × 2 SFBC for full buffer traffic*

Scenario	Sector spectral efficiency (bps/Hz)	Cell-edge spectral efficiency (bps/Hz)
Indoor hotspot	2.61	0.189
Urban micro-cell	1.69	0.040
Urban macro-cell	1.19	0.022
Rural macro-cell	1.30	0.021

3.9 Rel-8 interference coordination schemes

Two types of downlink inter-cell interference coordination (ICIC) schemes are supported in LTE Rel-8 – static and semi-static ICIC. In the static ICIC scheme (also known as fractional frequency reuse (FFR)), a portion of the bandwidth in each sector or eNB of the same physical site is reserved for scheduling cell-edge UEs. As an example, each sector reserves 18 resource blocks for scheduling of cell-edge users but uses only 6 orthogonal resource blocks (corresponding to a reuse pattern of 3). Since different cells use different sets of resource blocks, there is no interference with cell-edge users from sectors within the same physical sites. This can significantly improve the SINR of the cell-edge users at the expense of resource blocks. There are various FFR schemes. In a variant known as soft FFR, one can assign high-SINR UEs in the FFR zones used by neighboring sectors but with the eNB using a lower power level to these UEs. The static ICIC scheme needs some level of system planning. System simulations were run for full buffer traffic, for SU-MIMO with four transmit and two receive antennas, using a wideband scheduler and using an optimized FFR scheme. It was observed that the fifth-percentile edge throughput improved by approximately 11% while the sector throughput dropped by approximately 10%.

In semi-static ICIC, the amount of resources reserved for cell-edge users is semi-statically changed using an event-triggered message over the X2 interface. This reduces the system-planning impact. In LTE Rel-8, the Relative Narrowband Transmit Power (RNTP) message is defined for

Table 3.16. *Similarities and differences between LTE TDD and FDD*

Features	LTE FDD	LTE TDD	Comments
Frame structure	1-ms subframe	1-ms subframe	1-ms special subframe consisting of DwPTS, UpPTS, and GP
Switching points	–	5- and 10-ms periodicity	
eNB synchronization	Asynchronous/ synchronous	Synchronous	
DL control channel	Can schedule one DL and one UL subframe at a time	Can schedule one DL and multiple UL subframes at a time	
UL control channel	Single ACK/ NACK corresponding to one DL subframe	ACK/NACK bundling and multiple ACK/NACK corresponding to multiple DL subframes	

Table 3.16. (*cont.*)

Features	LTE FDD	LTE TDD	Comments
Random access channel	Four PRACH formats	Five PRACH formats	Short RACH transmitted on UpPTS
Special slot usage	–	DwPTS: RS, data and control UpPTS: SRS and short RACH	
HARQ timing	$N = 8$ stop-and-wait protocol	Dependent on DL:UL split	Affects performance of real-time services for TDD

DL, downlink; UL, uplink.

this purpose. It is transmitted from each cell when the transmit power of a set of resource blocks exceeds a specified threshold. The frequency of the RNTP transmission is limited to no more than 200 ms in order to prevent messaging overload.

3.10 LTE FDD vs. TDD comparison

As noted earlier in the chapter, there is very close alignment in the physical-layer parameters and features between FDD and TDD. A short summary of similarities and differences between LTE TDD and FDD is outlined in Table 3.16.

References

[1] Classon, B., Baum, K., Nangia, V. *et al.*, "Overview of UMTS air-interface evolution," *IEEE 64th Vehicular Technology Conference*, September 2006.

[2] Furuskar, A., Jonsson, T., Lundevall, M., "The LTE radio interface – key characteristics and performance," *IEEE 19th International Symposium on Personal, Indoor and Mobile Radio Communications*, September 2008.

[3] 3GPP TS 25.814, Physical layer aspects for evolved Universal Terrestrial Radio Access (UTRA), v7.1.0, September 2006.

[4] Kian, C. B., Doufexi, A., Armour, S., "Performance evaluation of hybrid ARQ schemes of 3GPP LTE OFDMA system," *IEEE 18th International Symposium on Personal, Indoor and Mobile Radio Communications*, September 2007.

[5] Yong, F., Lunden, P., Kuusela, M., Valkama, M., "Efficient semi-persistent scheduling for VoIP on EUTRA downlink," *IEEE 68th Vehicular Technology Conference*, September 2008.

[6] 3GPP TS 36.213, Physical layer procedures, v8.8.0, September 2009.

[7] Love, R., Kuchibhotla, R., Ghosh, A. *et al.*, "Downlink control channel design for 3GPP LTE," *IEEE Wireless Communications and Networking Conference*, pp. 813–818, April 2008.

[8] Liu, J., Love, R., Stewart, K., Buckley, M. E., "Design and analysis of LTE physical downlink control channel," *IEEE 69th Vehicular Technology Conference*, April 2009.

[9] Ghosh, A., Ratasuk, R., Frank, C. *et al.*, "Control channel design for high speed downlink shared channel for 3GPP W-CDMA, Rel-5," *IEEE 57th Vehicular Technology Conference*, vol. 3, pp. 2085–2089, April 2003.

[10] Hosein, P., "Resource allocation for the LTE Physical Downlink Control Channel," *IEEE GLOBECOM Workshops*, December 2009.

[11] Yang, Y., Che, W., Yan, N., Tan, X., Min, H., "Efficient implementation of primary synchronisation signal detection in 3GPP LTE downlink," *Electronics Letters*, vol. 46, no. 5, pp. 376–377, March 2010.

[12] Yingming, T., Guodong, Z., Grieco, D., Ozluturk, F., "Cell search in 3GPP Long Term Evolution systems," *IEEE Vehicular Technology Magazine*, vol. 2, no. 2, pp. 23–29, June 2007.

[13] Sanchez, J. J., Morales-Jimenez, D., Gomez, G., Enbrambasaguas, J. T., "Physical layer performance of Long Term Evolution cellular technology," *16th IST Mobile and Wireless Communications Summit*, July 2007.

[14] ITU-R M.2135, Guidelines for evaluation of radio interface technologies for IMT-Advanced, 2008.

[15] Yakun, S., Xiao, W., Love, R. *et al.*, "Multi-user scheduling for OFDM downlink with limited feedback for evolved UTRA," *IEEE 64th Vehicular Technology Conference*, September 2006.

[16] Farajidana, A., Chen, W., Damnjanovic, A. *et al.*, "3GPP LTE downlink system performance," *IEEE Global Telecommunications Conference*, November–December, 2009.

[17] Rinne, M., Kuusela, M., Tuomaala, E. *et al.*, "A performance summary of the evolved 3G (E-UTRA) for voice over Internet and best effort traffic," *IEEE Transactions on Vehicular Technology*, vol. 58, no. 7, pp. 3661–3673, September 2009.

[18] Berkmann, J., Carbonelli, C., Dietrich, F., Drewes, C., Wen, X., "On 3G LTE terminal implementation – standard, algorithms, complexities and challenges," *International Wireless Communications and Mobile Computing Conference*, pp. 970–975, August 2008.

[19] Puttonen, J., Henttonen, T., Kolehmainen, N. *et al.*, "Voice-over-IP performance in UTRA Long Term Evolution downlink," *IEEE Vehicular Technology Conference*, pp. 2502–2506, May 2008.

4　Uplink transmission and system performance

4.1 Introduction

LTE uplink provides an increase in capacity (both sector and cell-edge) by a factor of 2–3 compared with previous UMTS high-speed uplink packet access (HSUPA) systems at substantially less latency. This enables efficient support of high-rate data services such as FTP, HDTV broadcast, and HTTP as well as delay-sensitive services such as VoIP and video streaming. In LTE, several technological enhancements have been introduced in the uplink air interface to enable this improvement. They include orthogonal uplink transmission from intra-cell users, frequency-selective scheduling, shorter subframe size, support for 64-QAM modulation, multi-user spatial multiplexing, subframe bundling, semi-persistent scheduling, fractional power control, inter-cell interference control, and efficient control channels.

In this chapter, a detailed overview of LTE uplink structure, operations, and performance is provided. First, a description of the uplink structure, including the transmission scheme, frame structure, and physical channels, is given. Then, data- and control-channel operations supporting various uplink improvements such as orthogonal uplink transmission from intra-cell users, frequency-selective scheduling, subframe bundling, and semi-persistent scheduling are described. This is followed by an explanation of the reference signals, random access, and timing advance needed for uplink operations. Power-control methods are next described. Finally, details of the link- and system-level performance of the uplink are provided.

4.2 Transmission scheme and frame structure

The uplink physical layer provides three basic functions – data transport from the UE to the eNB, channel-state and control information feedback, and random access. In the uplink, three physical-layer channels are present – the Physical Uplink Shared Channel (PUSCH), Physical Uplink Control Channel (PUCCH), and Physical Random Access Channel (PRACH) – to support these functionalities. The PUSCH carries data traffic and is assigned to users via the uplink scheduling assignment. The PUCCH carries three types of control signaling – ACK/NACK for downlink transmission, scheduling request indicators, and feedback of downlink channel information, including the channel-quality indicator, precoding-matrix indicator, and rank indicator. The precoding-matrix and rank indicators provide information about the channel state needed for downlink MIMO transmission. The PRACH carries the random-access channel. Users may use this channel to request initial access, initiate handoff procedures, and transition from the idle to the connected state. In addition to the three physical-layer channels, two types of reference signals are supported on the uplink – the demodulation reference signal (DMRS), which is associated with transmission of uplink data or control, and the sounding reference signal (SRS), which is used mainly to aid in channel-dependent scheduling, timing maintenance, and UE-specific beamforming. Table 4.1 summarizes the different types of uplink physical channels and signals.

A graphical illustration of the uplink frame structure is shown in Figure 4.1. The uplink subframe structure is the same as that for the downlink and is 1 ms long and divided into two 0.5-ms slots. Each slot is comprised of either seven SC-FDMA symbols, in the case of a normal cyclic prefix, or six symbols, in the case of an extended cyclic prefix. Data transmission occurs in the inner band resource blocks in order to reduce out-of-band emission. Different users are assigned different resource blocks, ensuring orthogonality among users in the same cell. Data transmission may hop at the slot boundary to provide frequency diversity. Control resources are then placed at the edge of the carrier band, with inter-slot hopping to provide frequency diversity. The reference signals necessary for data demodulation are interspersed throughout the data and control channels. The SRS can be scheduled by the base station to be transmitted in the last symbol of a subframe.

A physical resource block consists of $N_{\text{symb}}^{\text{UL}} \times N_{\text{sc}}^{\text{RB}}$ resource elements, where $N_{\text{sc}}^{\text{RB}} = 12$ is the number of resource elements (or subcarriers) per

Table 4.1. *Summary of uplink channels and signals*

Name	Description
Physical channels	
Physical Uplink Shared Channel (PUSCH)	Shared data channel used for transmission of scheduled user data
Physical Uplink Control Channel (PUCCH)	Control channel used for transmission of control information including acknowledgments, scheduling requests, and downlink channel reports (channel quality, preferred precoding-matrix indicator, and rank indication)
Physical Random Access Channel (PRACH)	Random access used for transmission of random-access preambles
Physical signals	
Demodulation reference signal (DMRS)	Reference signal used for demodulation of data and control information
Sounding reference signal (SRS)	Reference signal used for sounding of the uplink channel response

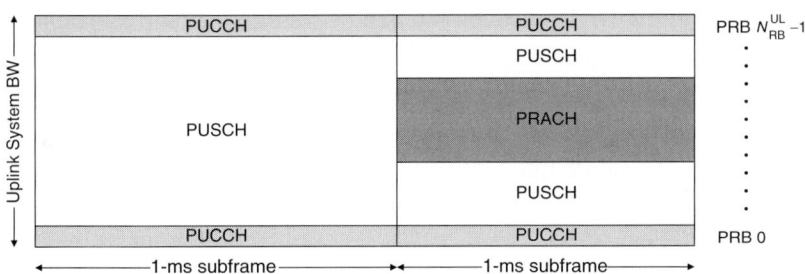

Figure 4.1. Uplink frame structure.

SC-FDMA symbol and $N_{\mathrm{symb}}^{\mathrm{UL}}$ is the number of symbols in a slot. The parameter $N_{\mathrm{symb}}^{\mathrm{UL}}$ equals 7 for a normal cyclic-prefix length and 6 for an extended cyclic-prefix length. The resource blocks are numbered $0, \ldots, N_{\mathrm{RB}}^{\mathrm{UL}} - 1$ and mapped in the frequency domain as shown in Figure 4.1.

4.3 Data channel

Data transmission occurs on the PUSCH. Users are allocated resources within the PUSCH by the eNB via a scheduling assignment. The allocation is given in multiples of virtual resource blocks. Each virtual resource block is mapped to two physical resource blocks. The size of each physical resource block is 12 resource elements by 7 SC-FDMA symbols for a normal cyclic prefix. However, one symbol per slot is used for the reference signal. Thus, each virtual resource block can accommodate 144 data symbols. For an extended cyclic prefix, the size of each physical resource block is 12 resource elements by 6 SC-FDMA symbols, resulting in 120 available data symbols after reference signals overhead. In LTE Rel-8, SU-MIMO is not supported since terminals can only transmit using one antenna. However, MU-MIMO is supported. With MU-MIMO, several users may be assigned the same resource blocks simultaneously on the same uplink subframe, and these overlapping resource blocks will be separated at the eNB using signal-processing techniques.

The transmitter chain can be summarized as follows. First, user data is coded and rate-matched to the assigned coding rate. The coded data is next scrambled on the basis of user and cell identities, modulated, DFT precoded, and then mapped to resource elements. For an efficient implementation of the DFT precoder, the number of assigned virtual resource blocks must be equal to $2^a 3^b 5^c$ where a, b, and c are non-negative integers. This means that the number of assigned resource blocks must be a multiple of 2, 3, or 5. To efficiently support small data transmission, allocation of resource blocks of size 1 is also allowed. For instance, at 20 MHz, where 100 virtual resource blocks are available, only 34 different resource-block assignments (1, 2, 3, 4, 5, 6, 8, 9, 10, 12, 15, 16, 18, 20, 24, 25, 27, 30, 32, 36, 40, 45, 48, 50, 54, 60, 64, 72, 75, 80, 81, 90, 96, 100) are possible. Although the granularity of the resource-block assignment is significantly reduced, it was shown in [1] that this resulted in only a small degradation (1%–2% capacity loss) of performance.

To preserve the single-carrier property in the uplink, only one physical channel can be transmitted in one SC-FDMA symbol. As a result, when data and control information coincide in the same subframe, control

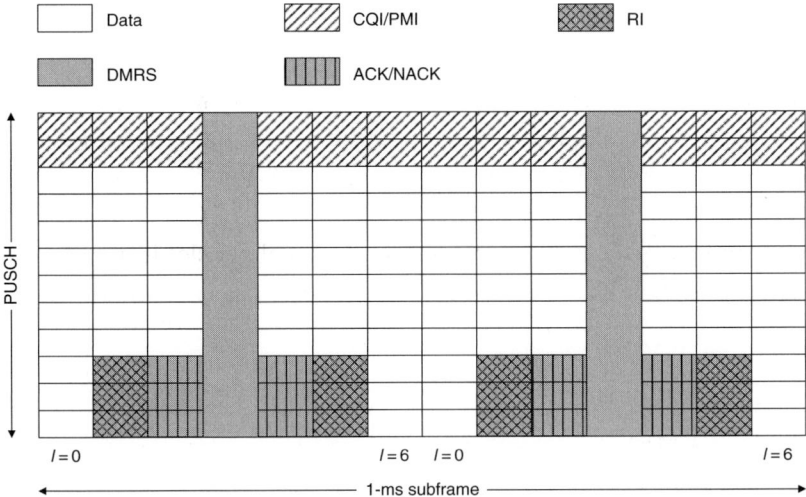

Figure 4.2. Data and control multiplexing on the PUSCH.

information must be multiplexed into unique positions in the PUSCH [2]. In this case, coded data is rate-matched or punctured to accommodate the additional control information. This allows the shared data, CQI/PMI, ACK/NACK, and rank indication (RI) to be transmitted in the same subframe. Figure 4.2 illustrates the multiplexing of data and control from the same user within one PUSCH data transmission. ACK/NACK and RI control information are transmitted next to the reference signal to provide the best channel-estimation performance. CQI is accommodated within the PUSCH allocation by appropriate rate-matching (i.e. adjusting the code rate) of the data.

Since different types of control information and data have different performance requirements, the number of PUSCH symbols used for each type of control may vary. As a result, users are configured with different control-information offset values that allow the size of each control region shown in Figure 4.2 to vary depending on the SNR requirements for each type of control information.

A synchronous HARQ protocol is used in the uplink. This means that retransmission for a specific HARQ process occurs on the basis of predefined timing relationships. For FDD, retransmission for an uplink packet transmitted at subframe n occurs in subframes $n + 8k$, where k is a positive integer. In addition, packet retransmission may be adaptive or non-adaptive. With non-adaptive retransmission, the UE retransmits the packet using the same frequency resource allocation as assigned previously. However, because incremental redundancy is used, the redundancy version will change. The redundancy version is implicitly tied to the transmission number and is given by 0, 2, 3, and 1. The pattern repeats if more transmission is needed. The advantage of this method is that an uplink scheduling grant is not required for retransmission, thus saving significant PDCCH overhead. With adaptive retransmission, the UE may be reassigned uplink frequency resources, modulation and coding rate, and redundancy version via a new uplink scheduling grant. This allows the eNB full flexibility in packet retransmission at a cost of additional PDCCH overhead.

4.3.1 Dynamic uplink scheduling assignment

Uplink data transmission can be scheduled dynamically every subframe or semi-persistently until the resource is released. For dynamically scheduled data transmission, the scheduling assignment is given via downlink control information (DCI) format 0, which is transmitted in the downlink control channel. In the uplink, synchronous HARQ is used and therefore retransmission will occur with predefined timing. As a result, it is not necessary to give another scheduling assignment for retransmission. In the absence of a scheduling grant, the user will retransmit the data packet using scheduling information obtained in the original assignment but with a different redundancy version. The eNB, however, can provide a scheduling assignment for retransmission, allowing it to change resource assignment, modulation and coding rate, and redundancy version. Table 4.2 lists the fields contained in DCI format 0. To ensure that the user can determine whether the DCI is decoded correctly, a 16-bit cyclic redundancy check is added. In addition, since each DCI is directed to a particular user, a 16-bit user identity given by the Radio Network Temporary ID (RNTI) is also embedded. To reduce overhead, however, the CRC is masked by the user identity and only one 16-bit field is

Table 4.2. *Uplink scheduling assignment (DCI format 0)*

Field	Size (bits)
Format differential flag	1
Hopping flag	1
Resource assignment	$\lceil \log_2 (N_{RB}^{UL}(N_{RB}^{UL} + 1)/2) \rceil$
Modulation and coding scheme	5
New data indicator	1
Transmit power control	2
Reference signal cyclic-shift value	3
Uplink index	2
Downlink assignment index	2
CQI request	1
CRC + UE ID	16

required, saving approximately 25%. This masking operation, however, results in a slight degradation in error-detection performance. Upon successful reception and decoding of the DCI, the user will transmit uplink data $n \geq 3$ subframes later. In FDD, $n = 3$, but for TDD the value of n is dependent on the TDD configuration and downlink subframe for which the scheduling assignment was received.

In LTE, different formats are distinguished by the users through differences in the size of the control information. Users are configured to receive several possible DCI formats depending on the desired transmission mode. Upon successful decoding of a control channel, the UE will determine the appropriate DCI format through the size of the DCI. However, DCI formats 0 and 1A have identical size to reduce blind decoding attempts by the user. Therefore, a format differential flag is present to allow the user to distinguish between these two formats. The hopping flag informs the user whether PUSCH frequency hopping is to be performed. When hopping is enabled, PUSCH transmission will hop within a subframe on a slot basis to provide greater frequency diversity. Two hopping modes are defined, and UEs are pre-configured in one of the two hopping modes. In Type 1 hopping mode, PUSCH transmission will hop according to information given in the scheduling assignment. This gives the eNB full flexibility in resource assignment to this particular user, and allows the eNB to resolve any potential resource conflict (e.g. when the PRACH is present). In Type 2 hopping mode, PUSCH transmission will hop according to a predefined pattern. This allows simple application of PUSCH frequency hopping at the loss of some scheduling flexibility.

The resource-assignment field provides resource-block allocation, and its size is dependent on the uplink bandwidth. For example, for a system bandwidth of 10 MHz, the length of this field is 11 bits, whereas, at 20 MHz, it is 13 bits. Since LTE requires assigned resource for a user to be contiguous in order to preserve the single-carrier property, the resource-assignment field is comprised simply of a starting virtual resource block number and length in number of contiguous resource blocks. When frequency hopping is disabled, the actual physical resource blocks are the same as the assigned virtual resource blocks. When frequency hopping is enabled, however, the physical resource blocks are mapped from the assigned virtual resource blocks on the basis of a pre-configured hopping mode and subframe number.

The modulation and coding scheme (MCS) field provides the modulation and coding rate to be used for uplink data packets. QPSK and 16-QAM are always available for all terminal categories, with 64-QAM supported only by category-5 terminals. Only turbo coding is used for data transmission, and coding rates of ~0.14 to ~0.84 can be supported with rate lower than 1/3 obtained through repetition coding. This allows a spectral efficiency per resource element of 0.28 to 5.0. In the uplink, the redundancy version used for incremental redundancy is implicitly given as 0, 2, 3, 1 depending on the retransmission number. That is, the first transmission of a data packet is implicitly given a redundancy version of 0. Retransmission will then follow the predefined pattern unless an explicit scheduling assignment is given to change the redundancy version. The new data indicator field indicates whether this packet is a new data packet or a retransmission of previous data packets. It is toggled between 0 and 1 whenever a new data packet is transmitted by the user.

The transmit power control (TPC) command directs the user to adjust the transmit power of the data packet on the basis of the assigned value. This command is part of the closed-loop power-control operation, which may be used for tracking fading conditions or to correct measurement errors as necessary. The reference signal cyclic-shift field provides the cyclic-shift value for the demodulation reference signal sequence. It can be used to assigned unique sequences to users transmitting on the same resource blocks as part of MU-MIMO operation or to manage PHICH assignment. Eight possible values are available.

The uplink index and downlink assignment index fields are applicable only to TDD. In addition, the uplink index field is used only in TDD uplink–downlink configuration 0. In this configuration, more uplink than downlink subframes are available and this field is used to schedule up to two uplink subframes via one scheduling assignment. In downlink-heavy TDD configurations, the downlink assignment index is used to inform the user of the number of scheduling assignments it may receive. For example, in TDD configuration 2, four downlink subframes are present for every uplink subframe. In this case, the UE may receive up to four downlink scheduling assignments prior to having to transmit an acknowledgment in the uplink. The downlink assignment index can then be used to update the UE regarding the total number of assignments to be expected within this scheduling window. This allows the UE to possibly detect whether it has missed any assignment within the window and transmit the appropriate response. Finally, the CQI request field is used to inform the user to also send CQI and PMI reports in the allocated subframe. This allows the eNB to receive full and fresh CQI reports, which may then be used to schedule upcoming downlink data.

4.3.2 Semi-persistent uplink scheduling assignment

In LTE, uplink scheduling assignment can be done dynamically or in a semi-persistent manner. Dynamic scheduling assignment is valid only for one subframe, whereas for semi-persistently scheduled data transmission an uplink allocation is given and remains valid until the allocation is released. For instance, using semi-persistent assignment, a user can be scheduled to transmit uplink data every 20 ms until this assignment is cancelled. This mode is designed to save control-channel resources by periodically and persistently allocating uplink resources to users with known periodic uplink data transmission such as VoIP or video conferencing. Because the scheduling is given just once, this option is particularly useful for users in poor channel conditions who take up a disproportionate amount of control-channel resources. As a result, significantly more users can be scheduled in a subframe than otherwise can be accommodated using only dynamic scheduling assignment on the downlink control channel. Configuration of semi-persistent allocation is done via RRC and L1 PDCCH signaling. Two parameters are configured by RRC signaling – the semi-persistent scheduling interval and the number of empty transmissions before the resource is implicitly released. Semi-persistent scheduling intervals of 10, 20, 32, 40, 64, 80, 128, 160, 320, and 640 ms are available. The number of empty transmissions before the resource is implicitly released can be configured to be 2, 3, 4 or 8. In addition, a user is also assigned a semi-persistent scheduling RNTI (SPS C-RNTI) that will be used in conjunction with L1 PDCCH signaling via a specially configured DCI format 0 to activate and release semi-persistent scheduling assignment. For activation of semi-persistent scheduling, DCI format 0 is used together with CRC that is masked by the SPS C-RNTI instead of C-RNTI. In addition, some fields are set to known values to provide enhanced error-checking capability and decrease CRC falsing probability compared with using the SPS C-RNTI alone. In the case of semi-persistent scheduling activation, the TPC field is set to 00, the reference signal cyclic shift is set to 000, and the most significant bit of the MCS field set to 0. The user can then obtain required parameters such as resource assignment and MCS from the remaining fields of DCI format 0. Explicit release of semi-persistent allocation is performed in a similar manner by the eNB by setting the TPC field to 00, the reference signal cyclic-shift field to 000, the MCS field to 11111, and resource-block fields to all ones.

Note that SPS assignment is valid only for initial transmission. If retransmission of a packet is required, the eNB must issue an explicit dynamic uplink scheduling assignment to handle the retransmission. In this case, the scheduling assignment is addressed to the SPS CRNTI.

4.3.3 Subframe bundling

LTE system coverage is usually limited by the uplink since the transmission power of the UE is much smaller than that of the eNB. For delay-insensitive traffic, HARQ can be used to balance downlink and uplink coverage. By increasing the maximum number of HARQ retransmissions in the uplink, uplink coverage can be increased at the cost of a reduced data rate. For delay-sensitive traffic, however, the maximum number of HARQ retransmissions is limited to within the delay budget. This is because packets that are received with delays exceeding the delay budget are normally discarded. As a result, in the uplink, LTE allows a packet to be transmitted over four consecutive uplink subframes via subframe bundling. This allows increased coding as well as increased transmission energy over multiple subframes. With subframe bundling, uplink coverage can be significantly increased [3]. Acknowledgment from the eNB is sent after the last subframe of the bundle has been transmitted. If retransmission is required, the entire bundle is retransmitted. An example of subframe-bundling operation is shown in Figure 4.3 for a 12.2-kbps AMR VoIP service that is transmitted in one uplink resource block. Compared with single-subframe operation, the total UE transmission power is increased by a factor of 4. If four resource blocks are used, the effective coding rate for the entire bundle is 0.15, compared with 0.6 when subframe bundling is not used.

Subframe-bundling mode is UE-specific, meaning that it can be configured on each UE individually, and can be enabled via the *ttiBundling* flag using the RRC configuration. For TDD, subframe bundling can be configured only for TDD configurations 0, 1, and 6, for which at least four uplink subframes are available per radio frame.

4.3.4 HARQ processes

To minimize packet retransmission latency, the maximum number of HARQ processes should be as small as possible. This is because retransmission must wait until all HARQ processes have been exhausted, so the greater the number of HARQ processes, the larger the retransmission

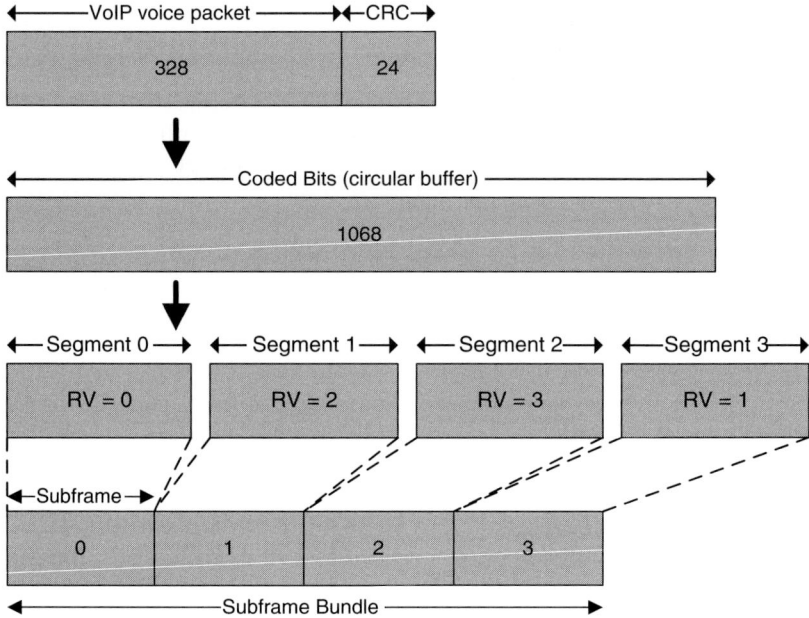

Figure 4.3. Example of subframe bundling operation.

delay. This number is therefore based on the maximum allowable processing time both at the eNB and at the UE. In LTE, this processing time should be less than 3 ms on the basis of an estimation of hardware capabilities. Using this processing time, Table 4.3 provides the maximum number of HARQ processes in the uplink. From this table, it can be seen that the retransmission latency is 7 ms for FDD and 9–10 ms for TDD. This provides a very quick turnaround in order to support low-latency services such as VoIP.

For FDD, a HARQ timing diagram for normal HARQ operation is shown in Figure 4.4. It can be seen that, using a processing time of 3 ms both at the eNB and at the UE, an uplink data transmission in subframe n will be acknowledged in subframe $n+4$, with earliest retransmission at subframe $n+8$. Because the retransmission can occur at the earliest on the eighth subframe following initial transmission as shown in Figure 4.4, eight different HARQ processes, each corresponding to a new data packet, can be supported.

Table 4.3. *Maximum number of uplink HARQ processes*

Frame structure type	Maximum number of HARQ processes		Retransmission latency (ms)	
	Normal	Subframe bundling	Normal	Subframe bundling
FDD	8	4	7	12
TDD 0	7	3	10	14
TDD 1	4	2	9	13
TDD 2	2	–	9	–
TDD 3	3	–	9	–
TDD 4	2	–	9	–
TDD 5	1	–	9	–
TDD 6	6	3	10	16

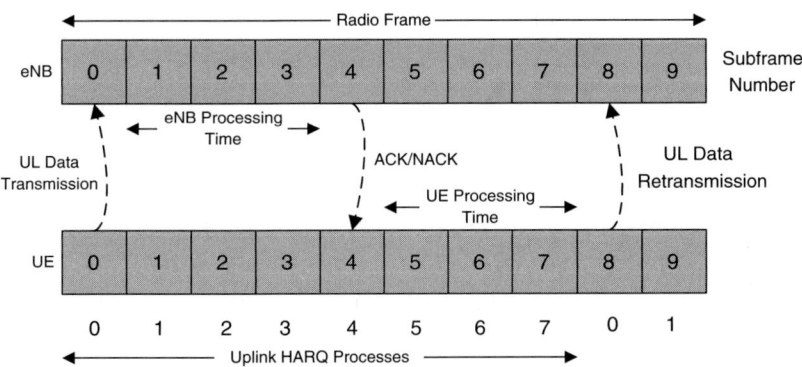

Figure 4.4. Uplink HARQ processes (FDD).

For TDD, the maximum number of HARQ process depends on the DL/UL configuration. This is because the eNB or UE may have to wait until a downlink or uplink subframe becomes available before transmission can occur. Figure 4.5 provides the HARQ timing diagram under normal operation for TDD configuration 0. It is seen that seven HARQ processes are available under normal operation and the packet retransmission latency is 10 ms. The values for other configurations are provided in Table 4.3.

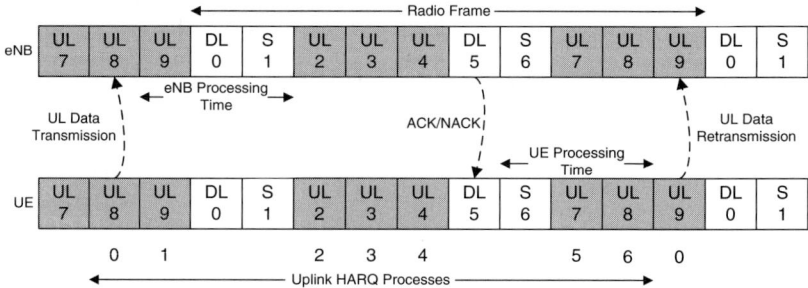

Figure 4.5. Uplink HARQ processes (TDD configuration 0).

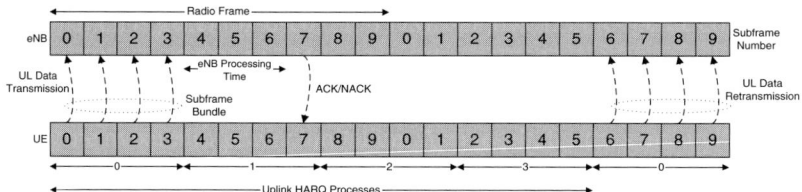

Figure 4.6. Uplink HARQ processes (FDD subframe bundling).

Under subframe-bundling operation, the maximum number of HARQ processes is determined by the bundle size of four subframes. A timing diagram for FDD is shown in Figure 4.6. In this example, four HARQ processes are available, with a retransmission time between bundles of 12 ms.

4.4 Control information

In the uplink, the following control information is conveyed by the mobile – ACK/NACK, scheduling request (SR), channel-quality information (CQI), precoding-matrix indicator (PMI), and rank indicator (RI) [4]. The ACK/NACK conveys packet acknowledgment and occurs in response to scheduled downlink data transmission. The SR, CQI, PMI, and RI are transmitted periodically in a predefined configuration by the base station. Users are allocated specific resources for their transmission.

For instance, a mobile may be instructed to report CQI information to the base station every 10 ms.

The ACK/NACK and SR are carried on the PUCCH, while CQI/PMI/RI reports are carried on either the PUCCH or the PUSCH. The PUCCH is placed on the band edge to allow contiguous data resource blocks and also to limit out-of-band emission from high-power data transmission. In addition, each PUCCH transmission is mapped into two slots that are located on opposite band edges to provide frequency diversity. An illustration of the PUCCH mapping to physical resource blocks is shown in Figure 4.7.

To preserve the single-carrier property in the uplink, only one physical channel can be transmitted in a subframe. As a result, when multiple items of control information must be sent simultaneously on the PUCCH, several formats are defined to handle different multiplexing options. Table 4.4 lists the various PUCCH formats available. The CQI/PMI/RI information is carried by PUCCH format 2/2a/2b while ACK/NACK and SR are carried by PUCCH format 1/1a/1b. The PUCCH format 2/2a/2b is mapped to the outermost control resource blocks (e.g. PUCCH 0 and 1 in Figure 4.7), while format 1/1a/1b is mapped to the inner control resource blocks (e.g. PUCCH 2 and 3 in Figure 4.7). This is because not all resource blocks reserved for PUCCH format 1/1a/1b may be used, and thus they can be reassigned to data transmission as they become available. The two formats are mapped to different resource blocks, with one transition resource block that may contain both PUCCH formats.

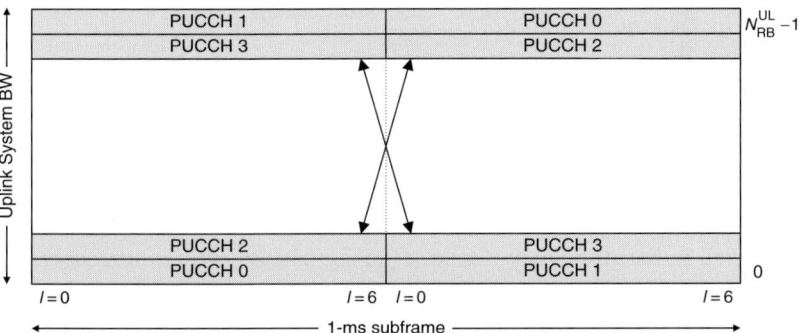

Figure 4.7. PUCCH mapping to physical resource blocks.

Table 4.4. *PUCCH formats*

PUCCH format	Field	Modulation	Number of information bits
1	Scheduling request	–	–
1a	1-bit ACK/NACK	BPSK	1
1b	2-bit ACK/NACK	QPSK	2
2	CQI/PMI/RI	QPSK	20
2a	CQI/PMI/RI + 1-bit ACK/NACK	QPSK + BPSK	21
2b	CQI/PMI/RI + 2-bit ACK/NACK	QPSK + QPSK	22

Table 4.5 summarizes the multiplexing options when different types of control information are to be transmitted in the same subframe. In general, ACK/NACK and SR are prioritized over CQI/PMI/RI transmission.

4.4.1 ACK/NACK and scheduling request

Acknowledgment and scheduling request information are carried on the PUCCH formats 1/1a/1b. The PUCCH format 1 is used to convey SR information, while formats 1a and 1b are used to convey 1-bit and 2-bit ACK/NACK, respectively. The 2-bit ACK/NACK is used to convey two acknowledgments, one per codeword, when SU-MIMO transmission on the downlink with two codewords is used. The information is first modulated and then multiplied by a cyclically shifted spreading sequence of length 12 resource elements. For SR, ON/OFF keying is used and the information is denoted by the presence or absence of the signal. For ACK/ NACK, BPSK and QPSK modulations are used for 1-bit and 2-bit ACK/ NACK, respectively. Subsequent to spreading by a cyclically shifted sequence in the frequency domain, the signals are further spread block- wise over multiple OFDM symbols per slot in the time domain. Note that the reference signal is also spread block-wise in the time domain. The number of users that may be multiplexed into a resource blocks depends

Table 4.5. *PUCCH multiplexing options*

Control information			
CQI/ PMI/RI	ACK/ NACK	SR	Multiplexing method
×	×	×	CQI is dropped, transmit only ACK/NACK + SR
×	×		If UE is configured to transmit ACK/NACK and CQI simultaneously, use PUCCH format 2a or 2b for a normal cyclic prefix, and PUCCH format 2 for an extended cyclic prefix; otherwise CQI is dropped and only ACK/NACK is transmitted
×		×	CQI dropped, only SRI transmitted, as shown above
	×	×	With negative SR, the UE will transmit the using the ACK/NACK resource; with positive SR, UE will transmit ACK/NACK information using the SRI resource

on the number of usable cyclic shifts and the block-wise spreading gain. The number of frequency-domain users that can be multiplexed via cyclic shifts is dependent on the channel delay spread, while the number of time-domain users is dependent on the Doppler spread. Resources in PUCCH formats 1/1a/1b are assigned using a PUCCH resource index $n_{PUCCH}^{(1)}$. This resource index can then be uniquely mapped to resource block number, orthogonal sequence index, and cyclic shift.

Figure 4.8 provides a block diagram of the PUCCH formats 1/1a/1b for a normal cyclic prefix. In the frequency domain, 12 different cyclic shifts are available, while block-wise spreading in the time domain is limited to 3 due to the reference signal. This allows a maximum of 36 users to be multiplexed in one resource block. In practice, however, the number of usable cyclic shifts is limited due to channel delay spread.

Scheduling request reporting is configured by higher layers and users are provided with periodic transmission opportunity during which to send the SR. Users are configured via three RRC parameters – *sr-PUCCH-ResourceIndex*, *sr-ConfigIndex*, and *dsr-TransMax*. The parameter

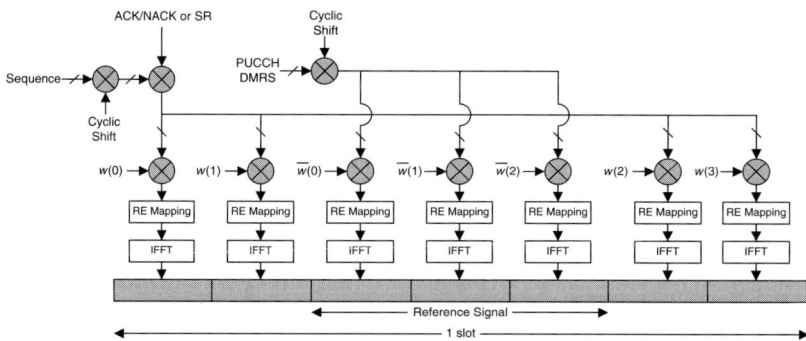

Figure 4.8. PUCCH formats 1/1a/1b channel structure.

sr-PUCCH-ResourceIndex provides the PUCCH resource index in which to transmit the SR. The parameter *sr-ConfigIndex* configures the reporting period and subframe offset for SR transmission. Valid periodicity values include 5, 10, 20, 40, and 80 ms. Finally, the parameter *dsr-TransMax* defines the maximum number of SR transmissions before the user will abandon the request. If the user does not receive a response to the SR after *dsr-TransMax* transmissions, it will assume loss of synchronization with the eNB and will initiate a random-access procedure.

The ACK/NACK transmission, on the other hand, is done in response to downlink data reception. In this case, a user selects the PUCCH resource to transmit the ACK/NACK on the basis of the type of scheduling assignment associated with downlink data reception. In the case of semi-persistently scheduled data, a user is pre-configured with four PUCCH resource indices and told which of the four pre-configured resource indices to use in the semi-persistent scheduling activation message. In the case of dynamically scheduled data, a user implicitly selects the PUCCH resource index according to the transmission mode.

In FDD, a user selects the PUCCH resource index on the basis of the index of the first CCE used in constructing the DCI and the number of reserved PUCCH resource indices. For example, if 20 PUCCH resource indices are reserved for RRC assignment (i.e. for SR and semi-persistently scheduled data) and the user were given a downlink scheduling assignment on CCEs 4, 5, 6, and 7, then the user will use PUCCH resource index $19 + 4 = 23$ to transmit the ACK/NACK.

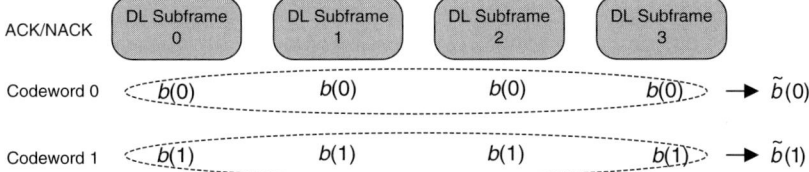

Figure 4.9. ACK/NACK bundling for TDD.

In TDD, for asymmetric DL/UL allocation, acknowledgments from multiple downlink subframes will have to be transmitted in one uplink subframe. For instance, in TDD configuration 2, in every radio frame eight downlink subframes are available versus two uplink subframes. As a result, each uplink subframe must carry the acknowledgments for four downlink subframes. As a result, two types of ACK/NACK feedback modes are available in TDD: bundling and multiplexing. In bundling mode, the user will bundle (i.e. perform a logical AND operation) all the acknowledgments for each codeword across subframes as shown in Figure 4.9. In essence, if any one of the data packets was not decoded correctly, then all transmissions will have to be retransmitted. This mode is useful when the UE is power-limited [5]. In this case, the user selects the PUCCH resource index on the basis of the index of the first CCE used in constructing the last correctly received DCI and the number of reserved PUCCH resource indices.

The ACK/NACK multiplexing mode is illustrated in Figure 4.10. In this mode, the number of feedback symbols is equivalent to the number of associated downlink subframes. For subframes without data transmission, the UE reports a DTX up to the eNB. To reduce the number of required feedback bits, spatial bundling across codewords is performed. That is, within each subframe, acknowledgments for the two MIMO codewords are bundled together to form one decision. For instance, in downlink subframe 0, if one codeword was received correctly while the other codeword was not, the decision for that subframe will be NACK. To transmit all the multiplexed bits, PUCCH resource selection is used in conjunction with a 2-bit feedback. The user selects the PUCCH resource index on the basis of the index of the first CCE used in constructing the

Table 4.6. *ACK/NACK multiplexing for two downlink subframes [27]*

Multiplexed ACK/NACK	PUCCH resource selection	Feedback bits
ACK, ACK	1	1, 1
ACK, NACK/DTX	0	0, 1
NACK/DTX, ACK	1	0, 0
NACK/DTX, NACK	1	1, 0
NACK, DTX	0	1, 0
DTX, DTX	–	–

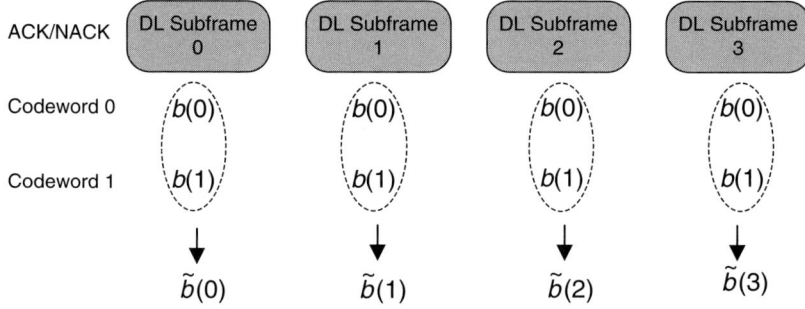

Figure 4.10. ACK/NACK multiplexing for TDD.

DCI, the feedback bits, and the number of reserved PUCCH resource indices.

An example of ACK/NACK multiplexing is shown in Table 4.6, where acknowledgments from two downlink subframes are to be multiplexed. By using PUCCH resource and symbol selection (i.e. how to set the feedback bits), six different combinations are supported.

Although the ACK/NACK channel is generally robust and performs well even for low SNR, the interference seen in this channel can be significant, since multiple users may be multiplexed into the same resource block. As a result, ACK/NACK repetition is supported in order to provide increased uplink coverage. Users are configured via the user-specific *ackNackRepetition* parameter. When ACK/NACK repetition is enabled, the user will transmit ACK/NACK with repetition factors

N_{ANRep} of 2, 4, and 6. For example, with $N_{ANRep} = 4$, the user will transmit the same acknowledgment in four consecutive subframes.

4.4.2 Channel measurement report – CQI/PMI/RI

In LTE, users may be configured to transmit downlink channel measurement reports including CQI, PMI, and RI back to the eNB to aid in scheduling, resource allocation, and link adaptation. The CQI report provides a measure of the supportable spectral efficiency (that is, the supportable modulation and coding rate) on the downlink shared data channel for the corresponding reporting bandwidth. Such CQI measurement reports play an integral part in system performance and are used in numerous system components such as scheduling, resource allocation, link adaptation, and downlink power allocation for control information. Similarly, the PMI report provides a preferred precoding codebook index for the corresponding reporting bandwidth. The PMI is used when closed-loop spatial multiplexing is used for downlink data transmission. The CQI and PMI reports can be wideband or per subband, which is defined as a set of contiguous resource blocks. The RI denotes the supported number of transmission layers or data streams as determined by the mobile. For instance, a rank indicator of two means that the mobile can support two independent data streams using MIMO technique.

Table 4.7 provides the supported 4-bit CQI values and their corresponding spectral efficiencies. For each CQI report, the user selects and reports the CQI index corresponding to the supportable modulation and coding rate (i.e. spectral efficiency) that it can receive with a transport-block error probability of less than 10%. For instance, when a user reports CQI index 8, it informs the eNB that, for the CQI bandwidth being reported, it can support a transport block using 16-QAM modulation and a coding rate of approximately 0.48 with a block error of less than 10%. Note that the CQI index is provided on the basis of the spectral efficiency supported by the mobile so as to provide a feedback mechanism that is independent of the performance at the mobile. However, from the observation that the supported spectral efficiency can be approximated by

Table 4.7. *The CQI table and reference SINR requirements [27]*

CQI index	Modulation	Spectral efficiency (bps/Hz)	Reference SINR (dB)
0	Out of range		
1	QPSK	0.15	−6.7
2	QPSK	0.23	−4.7
3	QPSK	0.38	−2.3
4	QPSK	0.60	0.2
5	QPSK	0.88	2.4
6	QPSK	1.18	4.3
7	16-QAM	1.48	5.9
8	16-QAM	1.91	8.1
9	16-QAM	2.41	10.3
10	64-QAM	2.73	11.7
11	64-QAM	3.32	14.1
12	64-QAM	3.90	16.3
13	64-QAM	4.52	18.7
14	64-QAM	5.12	21.0
15	64-QAM	5.55	22.7

$$\text{MPR} = \frac{1}{K_s} \log_2(1 + l \cdot \text{SINR})$$

the SINR required in order to support the assigned MPR is given by $\text{SINR} = (2^{\text{MPR} \cdot K_s} - 1)/l$ with typical values of $K_s = 1.25$ and $l = 0.66$. The approximate SINR corresponding to each CQI index is also provided for reference in Table 4.7. These approximate SINR values are not shown in the 3GPP specification, but are provided here for reference. Using this approximation, it is seen that the CQI index is given in approximately 2-dB step size. This provides enough resolution to cover an SINR dynamic range of approximately 30 dB.

The size of the CQI feedback depends on the reporting type as well as on the report mode, with possible values given in Table 4.8.

Table 4.8. *Size of CQI field*

Wideband CQI (bits)	4
Aperiodic subband differential CQI (bits)	2
Aperiodic subband differential CQI (bits)	3

Table 4.9. *Size of PMI field*

	Number of eNB antenna ports		
	2	2	4
Rank	1	2	≥ 1
PMI size (bits)	2	1	4

The PMI feedback is used when a mobile is configured in closed-loop spatial multiplexing mode, including single-layer spatial multiplexing (MU-MIMO and precoding) and multiple-layer spatial multiplexing (SU-MIMO). For other transmission modes, this reporting is not needed. In this mode, users are configured to report the preferred precoding entry from the relevant codebook using appropriate rank information. For instance, when only one-layer transmission can be supported ($RI = 1$), the user selects from codebook indices 0, 1, 2, and 3 when reporting the preferred PMI value to the eNB. Several codebook-index-selection criteria are available, and selection is implementation-specific to the particular mobile. Some well-known selection criteria include selecting the precoding entry that maximizes the SNR or capacity or minimizes the trace of the mean-squared error. The possible sizes of the PMI feedback, which is dependent on the number of transmit antenna ports at the eNB as well as on rank, are given in Table 4.9.

The RI feedback is used to inform the eNB of the number of transmission layers that can be supported by the mobile. This is usually determined by the rank of the composite channel response over all transmit and receive antennas, hence the name rank indicator. Rank determination can be done, for example, by eigen-decomposition of the composite channel response.

Table 4.10. *Size of RI field*

	Number of eNB antenna ports		
	2	4	4
Maximum number of layers	2	2	4
RI size (bits)	1	1	2

When a maximum of four transmission layers is supported, the 2-bit RI field has possible values of 1, 2, 3, and 4. However, when a maximum of two transmission layers is supported, the RI field is of size 1 bit with possible values of 1 and 2. The maximum number of supported transmission layers is given by the supported antenna ports at the eNB and UE. For example, to support four layers, four transmit antennas at the eNB and four receive antennas at the UE are needed. The eNB may determine the minimum number of receive antennas at the UE on the basis of UE category information exchanged at call initiation. The possible sizes of the RI feedback, which is dependent on the number of transmit antenna ports at the eNB as well as on rank, are given in Table 4.10.

4.4.2.1 CQI/PMI/RI reporting modes

Two reporting modes are supported for the downlink channel measurements – aperiodic reporting which is done on the PUSCH, and periodic reporting, which is done on the PUCCH. Aperiodic reporting is triggered by setting the CQI request bit in the uplink allocation. The user then sends the entire downlink channel report within the PUSCH allocation. This includes, for example, the wideband CQI value plus the CQI and PMI values for each subband, where a subband is defined as a set of contiguous resource blocks, and rank indication. Under periodic CQI reporting, the user is pre-configured via higher-layer signaling to transmit CQI reports periodically on the PUCCH. The user transmits one subband CQI/PMI report per reporting instance and cycles through the entire bandwidth. Both of these modes are optional, so the user may be configured in one or both modes. Alternatively, feedback need not be configured at all if the user can be supported without downlink channel state feedback.

In general, wideband feedback should be configured periodically to provide basic information about the downlink channel information to the eNB. In addition to this, narrowband feedback can be configured as needed to support frequency-selective scheduling and spatial multiplexing. The choice of periodic versus aperiodic narrowband reporting depends mostly on the downlink data traffic characteristics and overhead considerations. Naturally, the reporting mode should match the expected traffic pattern. For instance, if the mean time between downlink data traffic is long, aperiodic reporting can be used to supplement wideband feedback to reduce uplink overhead. On the other hand, for traffic with periodic transmission such as video conferencing, periodic reporting should be used.

Within each reporting mode, different types of reports can be configured as shown in Tables 4.11 and 4.12. For instance, in aperiodic reporting mode 1-2, the user will report one wideband CQI value and multiple PMI values (one per subband) when instructed to do so by the eNB. In aperiodic reporting, subband feedback may be UE-selected or higher-layer-configured. In the case of higher-layer-configured subband feedback, the user is configured to report feedback from a predetermined set of subbands. With UE-selected subband feedback, the user provides

Table 4.11. *CQI/PMI feedback types for aperiodic reporting [27]*

Aperiodic reporting mode	CQI	PMI
1-2	Wideband	Subband
2-0	UE-selected subband	None
2-2	UE-selected subband	UE-selected subband
3-0	Higher-layer configured subband	None
3-1	Higher-layer configured subband	Wideband

Table 4.12. *CQI/PMI feedback types for periodic reporting [27]*

Periodic reporting mode	CQI	PMI
1-0	Wideband	None
1-1	Wideband	Wideband
2-0	UE-selected subband	None
2-1	UE-selected subband	UE-selected subband

Table 4.13. *Supported feedback modes for PDSCH transmission mode [27]*

Transmission mode	Description of downlink data transmission	Aperiodic reporting mode	Periodic reporting mode
1	Single antenna	2-0, 3-0	1-2, 2-0
2	Transmit diversity	2-0, 3-0	1-2, 2-0
3	Transmit diversity if RI = 1, large-delay CDD for RI > 1	2-0, 3-0	1-2, 2-0
4	Closed-loop spatial multiplexing	1-2, 2-2, 3-1	1-1, 2-1
5	Multi-user MIMO	3-1	1-1, 2-1
6	Single-layer closed-loop spatial multiplexing	1-2, 2-2, 3-1	1-1, 2-1
7	Single antenna	2-0, 3-0	1-2, 2-0

feedback for a selected set of preferred subbands (for example, the best M subbands) together with positions of the preferred subbands.

The different types of reports are needed in order to support a wide range of downlink data transmission modes as shown in Table 4.13. For instance, when a user is configured in transmission mode 2 (transmit diversity), it may be configured to report downlink channel measurements using aperiodic reporting mode 2-0 or 3-0 and periodic reporting mode 1-2 or 2-0. On the other hand, when a user is configured in transmission mode 4 (closed-loop spatial multiplexing), PMI feedback is required. As a

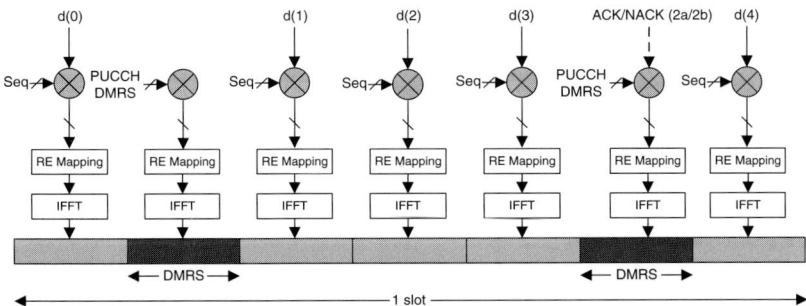

Figure 4.11. PUCCH formats 2/2a/2b.

result, the user must select from one of the feedback modes that include PMI feedback.

The block diagram for PUCCH formats 2/2a/2b used to carry periodic CQI/PMI/RI is shown in Figure 4.11. Note that only one slot of the subframe is shown. The CQI information bits (4–11 bits, depending on the reporting type) are encoded using Reed–Muller code into 20 coded bits. The coded bits are then mapped into 10 QPSK modulation symbols, and subsequently each modulation symbol is spread using a length-12 orthogonal sequence using the pre-configured cyclic shift. Twelve different cyclic shifts are available, leading to a CQI multiplexing capacity of 12 users per resource block. In practice, however, the maximum number of CQI reports per resource block is significantly smaller (e.g. 4–6) due to orthogonality and other cell interference. In formats 2a/2b, the second reference signal may be modulated by 1-bit or 2-bit acknowledgment when both items of control information are to be transmitted in the same subframe.

4.4.2.1.1 *Aperiodic CQI/PMI/RI reporting*

In aperiodic reporting, the feedback mode for a user is first configured via a higher layer according to Table 4.11. When downlink channel state information is needed at the eNB, it triggers the report by setting the CQI request flag in the uplink scheduling assignment. The user then transmits the downlink channel state information back to the eNB on the assigned PUSCH data resource. From Table 4.11, it can be seen that, for all modes, CQI or PMI feedback per subband is reported. The number of

Table 4.14. *Subband size for higher-layer-configured aperiodic reporting*

System bandwidth (MHz)	Number of resource blocks	Higher-layer-configured		UE-selected		
		Subband size (resource blocks)	Number of subbands	Subband size (resource blocks)	Number of preferred subbands	Number of subbands
1.4	6	–	1	–	1	1
3	15	4	4	2	3	8
5	25	4	7	2	5	13
10	50	6	9	3	5	17
15	75	8	10	4	6	19
20	100	8	13	4	6	25

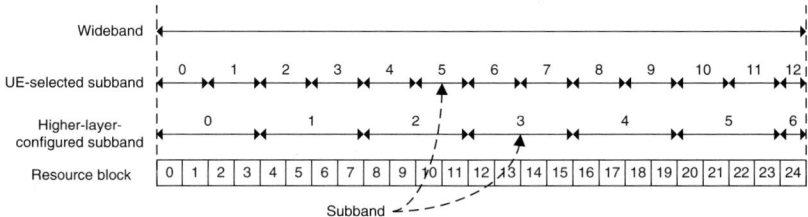

Figure 4.12. Wideband and subband size for aperiodic reporting.

subband-specific feedback reports depends on the subband size, which may be UE-selected or higher-layer-configured. In the case of higher-layer-configured subband feedback, the user is configured to report feedback from a predetermined set of subbands. With UE-selected subband feedback, the user provides feedback for a selected set of preferred subbands (for example, the best M subbands) together with positions of the preferred subbands. The subband size and number for each channel bandwidth are shown in Table 4.14. An illustration of how the channel is divided into subbands for UE-selected and higher-layer-configured subband reporting is shown in Figure 4.12 for a channel bandwidth of 5 MHz. In this case, for higher-layer-configured subband reporting, each subband is comprised of four contiguous resource blocks, with the exception of the last subband in the set, which is of size one resource block. Seven different subbands are available. Note that the wideband report is based on measurements over the entire channel bandwidth.

The following CQI/PMI reports are transmitted in each mode.

- *Mode 1-2: wideband CQI, subband PMI.* One preferred PMI for each subband using higher-layer-configured subband size. One wideband CQI report per codeword calculated using the preferred PMI in each subband. The user will transmit four (wideband CQI) bits per codeword. One codeword is assumed when the rank is 1 and two codewords are assumed when the rank is greater than 1. The size of the wideband PMI report is dependent both on the rank and on the number of antenna ports at the eNB due to the different codebook sizes. When four antenna ports are present at the eNB, the PMI report is 4 bits regardless of rank. However, when two antenna ports are present at the eNB, the PMI report is 2 bits for rank 1 and 1 bit for rank 2. In this case, the user will transmit N PMI reports, where N is the number of subbands. Using the example in Figure 4.12, in this mode a rank-2 user will transmit two wideband reports (one per codeword) and seven subband PMI reports.

- *Mode 2-0: UE-selected subband CQI, no PMI.* One wideband CQI report, one UE-selected subband CQI report, and the position of the selected subbands. The wideband report is calculated on the basis of the first codeword. The UE selects M preferred subbands from the set, where M is given in Table 4.14, and reports the CQI on the basis of transmission over the selected subbands, calculated on the basis of the first codeword. The UE-selected subband CQI is differentially encoded with respect to the wideband index. The valid offset values for the 2-bit differential field are ≤ -1, 2, 3, and ≥ 4. The position of the M preferred subbands is given by an L-bit index, where

$$L = \left\lceil \log_2 \binom{N}{M} \right\rceil$$

and N is the number of subbands. For instance, using the example in Figure 4.12, the UE will report one wideband report (4 bits), one UE-selected differential subband report (2 bits), and the position of the selected subbands ($L = 11$ bits with $N = 13$ and $M = 5$ from Table 4.14). Thus, the user will transmit a total of 17 bits in this example.

- *Mode 2-2: UE-selected subband CQI, UE-selected subband PMI.* One wideband CQI report per codeword, one UE-selected subband CQI report per codeword, one wideband PMI report, one UE-selected subband PMI report, and the position of the selected subbands. Similarly to mode 2-1, UE selects M preferred subbands from the set, where M is given in Table 4.14, and reports the position of the M preferred subbands using an L-bit index. The UE-selected subband CQI is likewise differentially encoded with respect to the wideband index. For instance, using the example in Figure 4.12 with four antenna ports and rank 1, the UE will report one wideband CQI report (4 bits), one UE-selected differential subband CQI report (2 bits), one wideband PMI report (4 bits), one UE-selected subband report (4 bits), and the position of the selected subbands ($L = 11$ bits with $N = 13$ and $M = 5$ from Table 4.14). Thus, the user will transmit a total of 25 bits in this example.

- *Mode 3-0: UE-selected subband CQI, no PMI.* One wideband CQI report and one differential CQI report per subband, calculated on the basis of the first codeword. The differential CQI value is with respect to the wideband index. The valid offset values for the 2-bit differential field are 0, 1, ≥ 2, and ≤ -1. For instance, with a reported wideband index of 7 and differential value ≥ 2, this subband can support transmission with CQI index ≥ 9. In general, the number of feedback bits for this mode is given by $4 + 2N$, where N is the number of subbands. Using the example in Figure 4.12, in this mode the user will transmit one wideband report (4 bits) and seven differential subband reports ($7 \times 2 = 14$ bits).

- *Mode 3-1: UE-selected subband CQI, wideband PMI.* For each codeword, one wideband CQI report and one differential CQI report per subband. In addition, a single wideband PMI is reported. Note that the size of the CQI reports is dependent on the rank. The user will transmit 4 (wideband CQI) + $2N$ (differential subband CQI) bits per codeword. One codeword is assumed when the rank is 1 and two codewords are assumed when the rank is greater than 1. On the other hand, the size of the wideband PMI report is dependent both on the rank and on the number of antenna ports at the eNB due to the different codebook sizes.

When four antenna ports are present at the eNB, the PMI report is 4 bits regardless of rank. However, when two antenna ports are present at the eNB, the PMI report is 2 bits for rank 1 and 1 bit for rank 2.

In addition to the above CQI/PMI reports, the RI is also reported if the user is configured in transmission mode 3 or 4 (open-loop or closed-loop SU-MIMO).

4.4.2.1.2   Periodic CQI/PMI/RI reporting

The following CQI/PMI reports are transmitted in each mode.

- *Mode 1-0: wideband CQI, no PMI.* One wideband CQI report per codeword calculated using the first codeword. For transmission mode 3 (open-loop spatial multiplexing), the rank is also reported, and the CQI is calculated on the basis of the last reported rank. The size of the report is 4 bits when the CQI is being reported. When the RI is being reported, the size of the report is 1 or 2 bits.

- *Mode 1-1: wideband CQI, wideband PMI.* One wideband PMI report, one wideband CQI report for the first codeword calculated using the wideband PMI report, and one 3-bit differential CQI report for the second codeword if the rank is greater than 1. The possible 3-bit differential CQI values are $0, 1, 2, \geq 3, \leq -4, -3, -2$, and -1, and represent the CQI offset level between values of the wideband CQI index for the first and second codewords. The size of the CQI report is 4 bits (wideband CQI for the first codeword) when the rank is 1 and 7 bits (wideband CQI for the first codeword, differential CQI for the second codeword) when the rank is greater than 1. The size of the PMI report, sent simultaneously with the CQI report, is given in Table 4.9. When the RI is being reported, the size of the repot is 1 or 2 bits.

- *Mode 2-0: UE-selected subband CQI, no PMI.* One wideband CQI report or one UE-selected subband report. The two types of report are sent depending on the configured reporting instance of each report. The wideband CQI report is calculated using the first codeword. For a UE-selected subband report, the UE selects one preferred subband in each of the J bandwidth parts and reports a CQI calculated on the basis of transmission in that subband alone. The user also reports the position of the selected subband using an L-bit index. The user will cycle through all

J bandwidth parts in successive reporting instances. In this mode, the size of the CQI report is 4 bits when the wideband CQI is being reported and $4 + L$ bits when the UE-selected CQI is being reported. When the RI is being reported, the size of the report is 1 or 2 bits.

- *Mode 2-1: UE-selected subband CQI, wideband PMI.* One wideband PMI and one wideband CQI report in a subframe configured for wideband reporting, and one UE-selected CQI report in a subframe configured for UE-selected reporting. When wideband reporting is being done, the UE reports one wideband PMI report, one wideband CQI report for the first codeword calculated using the wideband PMI report, and one 3-bit differential CQI report for the second codeword if the rank is greater than 1. When UE-selected subband reporting is being done, the UE selects one preferred subband in each of the J bandwidth parts and reports a CQI calculated on the basis of transmission in that subband alone. When the rank is greater than 1, the UE also reports an additional 3-bit differential CQI report for the second codeword. The user also reports the position of the selected subband using an L-bit index. The user will cycle through all J bandwidth parts in successive reporting instances. When the RI is being reported, the size of the report is 1 or 2 bits. The size of the wideband CQI/PMI report is identical to that for mode 1-1 reports, while the size of the UE-selected subband CQI report is either $4 + L$ bits for rank 1 or $4 + L$ bits for rank greater than 1.

4.5 Reference signals

In the uplink, two types of uplink reference signals are present – the demodulation reference signal used for demodulation of the uplink data and control channels, and the sounding reference signal used for uplink channel sounding. The demodulation reference signal spans the bandwidth of the transmitted signal, while the sounding reference signal bandwidth is configurable. Both signals are constructed from the same set of constant-amplitude zero-auto-correlation (CAZAC) sequences. These CAZAC sequences exhibit the following important properties: low cubic metric, low cross-correlation between different sequences (either of the same length or of different lengths), large number of available sequences, and low sequence generation complexity and storage requirement. These properties make them attractive for use as reference signals. The reference signal sequence is given by

$$r_{u,v}^{(\alpha)} = e^{j\alpha n}\bar{r}_{u,v}(n)$$

where $\bar{r}_{u,v}(n)$ is the base sequence and α is the cyclic shift. The length of the sequences is in multiple resource blocks. The base sequence is uniquely identified by the group number u and the within-group number v. There are 30 different groups ($u = \{0, 1, \ldots, 29\}$). For sequences of length five resource blocks or more, there are two member sequences ($v = \{0, 1\}$) available within a group. Otherwise, only one member sequence ($v = \{0\}$) is available within a group. For sequences of length three resource blocks or more, extended Zadoff–Chu sequences, a type of CAZAC sequence, are used. Zadoff–Chu sequences have very nice properties, and can be generated via a formula that reduces the storage requirement. For smaller sequences, however, only a few low-cubic-metric Zadoff–Chu sequences are available. As a result, computer-generated CAZAC sequences are used instead.

In LTE, 30 different base sequence groups are available for assignment to different cells. This is limited by the number of available root Zadoff–Chu sequences in the case of of three resource blocks allocation. Within each cell, two different cell-specific interference-randomization techniques may be applied to the uplink reference signals – group and sequence hopping. Both techniques are used to alleviate large cross-correlations arising from Zadoff–Chu sequences of different lengths and ensure that interference from uplink reference signals in other cells is minimized. Group hopping entails random but coordinated selection of the group number u among different cells in order to change the sequence group number from slot to slot. This method can be used with simple cell planning of the sequence group since the group number in each cell will change with time. Alternatively, sequence hopping can be used instead of group hopping. Sequence hopping, which is available for sequence lengths larger than six resource blocks, randomly selects one of the two available base sequences on a slot-by-slot basis. This method can be used if a static sequence group associated with a cell is desired (e.g. due to cell planning based on an interference pattern).

4.5.1 Demodulation reference signal

The demodulation reference signal is used to demodulate the PUSCH and PUCCH channels. The demodulation reference signal is defined using four parameters – the sequence length, sequence group number, sequence number within group, and cyclic shift. For the PUSCH, the length of the signal is identical to the assigned bandwidth, while the sequence group number and sequence number within group are assigned as described in Section 4.5. There are 12 available cyclic shifts, which are configured by three component parameters – the cell-specific cyclic shift, the UE-specific cyclic shift given in the uplink assignment grant, and the random cyclic shift based on cell identity and slot number. Eight cell-specific cyclic-shift values are available, requiring three-bit signaling. The values are {0, 6, 3, 4, 2, 8, 10, 9} and are chosen for the minimum interference among cyclic-shift values. This allows sequence planning whereby different cell-specific shifts are assigned in order to minimize possible interference. Additionally, to ensure that interference is randomized (that is, the interference between two cells is random), a cell-specific random pattern is also imposed. In addition, eight UE-specific cyclic-shift values, {0, 2, 3, 4, 6, 8, 9, 10}, are available and dynamically assigned via the uplink scheduling assignment. This allows the eNB to uniquely identify the demodulation sequences from multiple users via different cyclic shifts in support of MU-MIMO where multiple users share the same resource blocks.

For the PUCCH, the demodulation reference signal spans only one resource block and the cyclic-shift value is dependent on the PUCCH format. The number of possible cyclic-shift values is configurable, with 12 as the maximum.

4.5.2 Sounding reference signal

The sounding reference signal (SRS) is used to sound the uplink channel, which allows the eNB to measure the uplink channel response. This allows the eNB to determine the channel-quality information in the uplink direction, perform accurate link adaptation, and support frequency-selective scheduling. It may also be used to beamform downlink data transmission on the basis of, for example, the angle of arrival or channel reciprocity property. In addition, the eNB may also use the timing information from the uplink channel response to maintain uplink synchronization. The sounding bandwidth, frequency position, periodicity, and subframe offset are configured by the eNB via higher-layer signaling on a cell-wide basis. Users are then configured on a per-user basis with different sounding periodicities, bandwidths, and hopping patterns based on the cell-wide configuration.

In LTE, the maximum sounding bandwidth is configured on a cell-wide basis. Eight SRS configurations (C_{SRS}) are available, with a specific value for each configuration dependent on the system bandwidth. The eight configurations provide the maximum SRS bandwidth. For example, for a system bandwidth of 10 MHz, SRS configurations of 16–48 resource blocks are available. LTE provides multiple user-specific sounding bandwidths to support various sounding strategies, requirements, and power limitations. This allows the operator to select the SRS bandwidth that is appropriate to the amount of control overhead and sounding strategy. For example, in interference-coordinated systems, users may be restricted to transmitting only in a certain resource-block subset. As a result, the sounding region can be configured to span only the resource blocks used. As another example, VoIP users may be restricted to transmission in certain groups of resource blocks, and thus do not require wideband sounding. Power limitation may force some users to utilize the narrow-band sounding mode. This is because, for power-limited users, the power spectral density decreases as the transmission bandwidth increases. Thus, to achieve the minimum received signal quality, the users need to restrict transmission of the SRS in narrowband operation. This situation is common for cell-edge users, who also suffer from high inter-cell interference in addition to possible power limitation.

Within each SRS bandwidth configuration, four different user-specific assignments (B_{SRS}) on the basis of a tree structure are possible. This allows efficient assignment of the SRS, especially when users with different sounding bandwidths are multiplexed together, because several code-assignment algorithms and strategies are available. Sequence planning and allocation can also be performed efficiently. An example is shown in

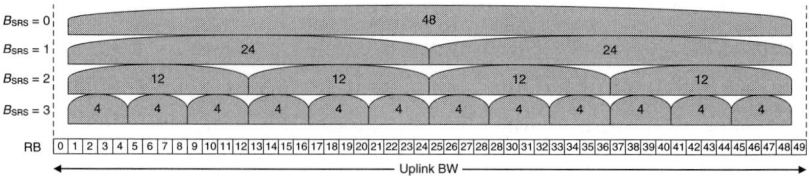

Figure 4.13. Sounding reference signal configuration example.

Figure 4.13 for $C_{\mathrm{SRS}} = 0$ for 10 MHz. In this case, the sounding region spans 48 resource blocks. The B_{SRS} assignment provides the sounding bandwidth for each sounding instance. For example, with $B_{\mathrm{SRS}} = 0$, the entire sounding region is sounded at once (wideband sounding). With $B_{\mathrm{SRS}} = 3$, only four resource blocks are sounded at a time (narrowband sounding), and users can be configured to hop among different sounding regions in each instance in order to eventually sound all 48 resource blocks.

Within each sounding region, a transmission comb is used with the sequence as given in Section 4.5 (i.e. the sequence used for the sounding is the same as the sequence used for the reference signal). The length of the sequence is determined by the sounding bandwidth. In LTE, a decimation factor of 2 is used, which means that only every other resource element is sounded and two transmission combs are available. Thus, to sound 4 resource blocks requires a sequence of length 24 resource elements. In addition, within each transmission bandwidth, eight cyclic shifts are available. Both FDM using a comb structure and CDM using cyclic shifts are used in LTE to increase the number of supportable users and provide orthogonality when different sounding bandwidths are used in the same symbol. Using Figure 4.13 as an example, two sounding bandwidths, say 48 and 4 resource blocks, can be configured in the same symbols using frequency-division multiplexing. One frequency comb can be configured to support an SRS transmission bandwidth of 48 resource blocks while the other comb can be configured to support an SRS transmission bandwidth of 4 resource blocks. This provides orthogonality among signals of different transmission bandwidths. In addition, within each transmission bandwidth, up to eight users can be code-division multiplexed via the use of different cyclic-shift values.

Consequently, a maximum number of 16 users can be multiplexed into one sounding region. However, this might not be achievable in practice due to orthogonality and interference issues. For example, at the smallest sounding bandwidth of four resource blocks, each cyclic shift provides timing protection of approximately 4.2 μs. In propagation channels with delay spread longer than this value, orthogonality is lost and a user will experience interference from other users using adjacent cyclic shifts. Additionally, with many users multiplexed into the same bandwidth, the SINR of the desired users may be degraded significantly due to other cell interference. As a result, a more realistic estimate of the number of supportable users may be approximately six to eight users in each sounding transmission. For the example shown in Figure 4.13, 6–8 users can be multiplexed into one sounding symbol with $B_{SRS} = 0$ and 72–96 users can be multiplexed into one sounding symbol with $B_{SRS} = 3$. Additional users may be time multiplexed on the basis of the required sounding period, which can significantly increase the number of users that can be supported. For instance, for low-mobility users, a sounding period in the range of 40–80 ms may be sufficient because the propagation channel changes slowly. This allows sounding to be efficiently supported even when many users are active in the system.

In FDD, sounding is performed by transmitting the configured sounding reference signal in the last SC-FDMA symbol of a subframe to provide the shortest possible delay between transmission and application of the SRS. As a result, users who are configured to transmit uplink data in a sounding subframe must puncture their uplink data transmission to accommodate sounding. In TDD, sounding can also be performed in the UpPTS in addition to in the data subframes. In both cases, the eNB configures the sounding period T_{SFC} and subframe offset Δ_{SFC} on a cell-wide basis. The possible numbers of sounding subframes for each 10-ms radio frame are $\{1, 2, 4, 5, 7, 8, 10\}$ for FDD and $\{2, 3, 4, 5, 6, 7, 8, 9\}$ for TDD. Once cell-wide sounding has been configured, the user-specific sounding period T_{SRS} and subframe offset T_{offset} can be individually configured. The available user-specific sounding period is $\{2, 5, 10, 20, 40, 80, 160, 320\}$ ms, with appropriate values of subframe offsets available depending on the period. An example sounding subframe configuration is shown in

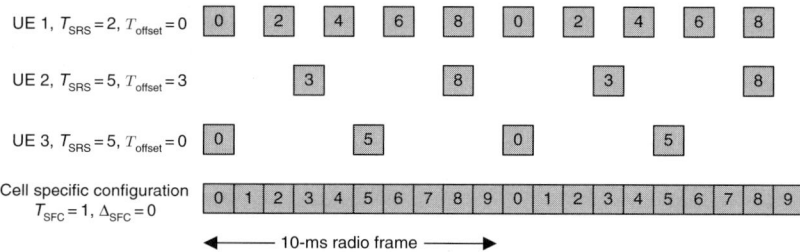

Figure 4.14. Sounding subframe configuration example – FDD.

Figure 4.14. In this example, cell-wide sounding is configured in every subframe and users are assigned different periods and offsets.

Note that the SRS transmission bandwidth does not extend into the PUCCH control region, so control information is not punctured. However, when a user has both control information and SRS to be transmitted in the same subframe, the following rules apply.

- SRS + PUCCH format 2/2a/2b. User will not transmit an SRS and transmits only PUCCH format 2/2a/2b.
- SRS + ACK/NACK or SR. User will transmit an SRS and transmit an ACK/NACK or SR only if the higher-layer parameter *Simultaneous-AN-and-SRS* is set to TRUE. In this case, a shortened PUCCH format is used to transmit the ACK/NACK or SR in order to preserve the single-carrier property. Otherwise, if the parameter *Simultaneous-AN-and-SRS* is set to FALSE, the user will not transmit an SRS and transmits only an ACK/NACK or SR.

In general, SRS has lower priority than other control signaling and as a result will usually be dropped when SRS transmission coincides with uplink control signals.

4.6 Random access

When the mobile is not time-synchronized to the base station in the uplink, it must use a contention-based random-access channel to access the network. It may use this channel to request initial access, initiate handoff procedures, and transition from idle to connected state. To ensure

Table 4.15. *Random-access preamble formats*

Preamble format	Cyclic-prefix length T_{CP} (µs)	Preamble sequence length T_{PRE} (µs)	Maximum supported cell size (km)	Subcarrier spacing (kHz)
0	103.125	800	14.6	1.25
1	684.375	800	101.8	1.25
2	203.125	2×800	29.7	1.25
3	684.375	2×800	101.8	1.25
4 (TDD only)	14.583	133.33	1.4	7.5

low latency, the random-access procedure must be designed such that a control-plane latency requirement of less than 100 ms is achieved. In LTE, five random-access formats are available, as shown in Table 4.15. Each PRACH occupies a bandwidth of six physical resource blocks (equivalent to a bandwidth of 1.08 MHz). In FDD mode, only one PRACH is available per subframe. However, its periodicity can be configured from one PRACH every 20 ms to one every 1 ms in order to provide sufficient random-access opportunities. The location and periodicity of the channels are signaled on the system information blocks. In TDD mode, due to the reduction in the number of available uplink subframes, multiple PRACHs may be defined in one subframe.

In LTE, random-access sequences (or preambles) must exhibit good detection performance and robustness with respect to interference as well as providing accurate timing estimation [6]–[7]. This is because, in E-UTRA, uplink transmissions must be synchronized in order to prevent interference. In addition, random access must be possible from the cell edge, where the SINR may be very poor. As a result, to meet the coverage requirement, only the preamble can be transmitted in the contention channel. This means that the user is able only to transmit a sequence but no additional data (e.g. user identity) for the initial random-access attempt. Figure 4.15 illustrates the random-access burst, which consists of a cyclic prefix, preamble sequence, and guard time. Table 4.15 lists the

Random Access

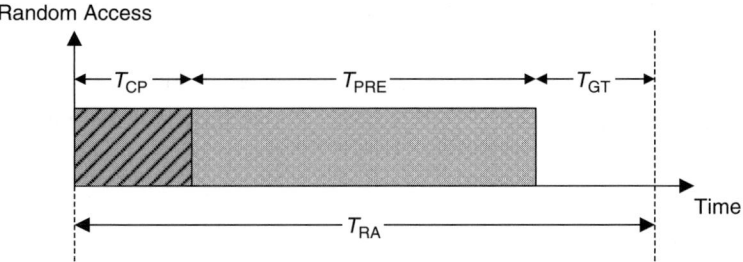

Figure 4.15. Random access burst.

available preamble formats and supportable cell sizes. In LTE, random access must be designed to support large cells of size up to 100 km. In this case, repetition is used to extend the random-access burst to allow more transmission energy. This requires appropriate adjustments to the cyclic prefix and guard period. For instance, to support a cell of size 25 km, an extended random-access burst of duration 2 ms is deployed as shown in Table 4.15, where the 800-µs preamble is repeated with the cyclic prefix length and guard interval each extended to 200 µs.

A cyclic prefix is added to aid in frequency-domain processing in order to reduce the detection complexity. A guard interval is required in order to prevent interference with other transmissions arising from timing misalignment due to different propagation delays. This timing misalignment between mobiles in the cell is dependent on the cell size. For instance, the guard interval 103.125 µs corresponds to a supportable cell radius of approximately 14.6 km. After removing approximately 5 µs to account for the propagation channel, the usable guard interval is approximately $(103.125\,\mu s - 5\,\mu s)/6.67\,\mu s/km \approx 14.6\,km$, where 6.67 µs/km accounts for twice the propagation delay between the base station and the mobile. Note that twice the propagation delay is accounted for because the mobile synchronizes to a delayed transmission of the base station.

Naturally, preamble waveforms for random access should have a good detection probability while maintaining a low false-alarm rate, allow accurate timing estimation, and have low power de-rating. In this case, a Zadoff–Chu sequence is used. A Zadoff–Chu sequence has the zero-correlation-zone property, which means that it has zero periodic cross-correlation for a contiguous set of delays. In other words, within a certain amount of delay, the cross-correlation of this sequence and its delayed version is zero. The constant amplitude results in a low peak-to-average power ratio in the transmitter. This is especially important in the uplink, where the peak-to-average power ratio must be kept low due to power-amplifier limitations. The Zadoff–Chu sequence of length N_{ZC} is given by the expression

$$x_u(n) = e^{-j\pi un(n+1)/N_{ZC}}, \quad 0 \le n \le N_{ZC} - 1$$

where u, the sequence index, is relatively prime to N_{ZC} (i.e. the only common divisor for u and N_{ZC} is 1). For a fixed u, the Zadoff–Chu sequence has ideal periodic auto-correlation property (i.e. the periodic auto-correlation is zero for all time shifts other than zero). For different values of u, Zadoff–Chu sequences are not orthogonal, but exhibit low constant cross-correlation regardless of time shift. If the sequence length N is selected as a prime number, there are $N_{ZC} - 1$ different available sequences. The zero-correlation zone for the Zadoff–Chu sequence is generated using the cyclic-shift version of the base carrier sequence as shown in Figure 4.16. Note that each zero-correlation zone must be large enough to accommodate the maximum timing misalignment between mobiles in the cell, which is dependent on the cell size. Thus, the number of zero-correlation zones that can be generated per sequence index u is based on the sequence length N_{ZC} and the cell size. The maximum available number of sequences available in the system is then ($N_{ZC} - 1$) $\times L$, where L is the number of zones per sequence index.

In LTE, two preamble sets are available – regular and restricted. The restricted set is used for high-speed cells where most users access the cell in a high-mobility environment (e.g. in high-speed trains). At high

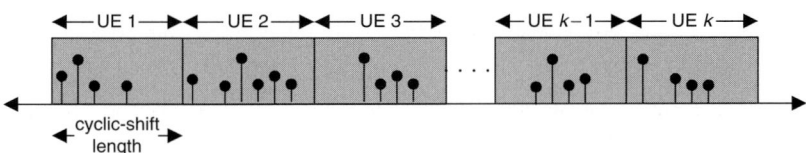

Figure 4.16. Preamble sequence for random access.

velocity, frequency offset due to the Doppler shift causes spurious or aliased peaks, resulting in high false-alarm rates. These spurious peaks occur at known cyclic-shift values, and, as a result, some cyclic shifts are restricted (i.e. not used) in the preamble set. This significantly improves the detection performance in a high-mobility environment at the cost of more Zadoff–Chu sequences being consumed in the construction of the preamble set.

4.6.1 Random-access procedure

Figure 4.17 provides the random-access procedure. Four different messages are exchanged as part of the random-access procedure and contention resolution. They are random-access preamble transmission, random-access response, L2/L3 message transmission such as connection request, and RRC contention resolution [8].

In the first step, a mobile randomly selects a PRACH and a preamble among the available preambles within the set. Two preamble groups (A and B) are available to be selected depending on the size of the L2/L3 message and pathloss to be transmitted by the user. In general, a user selects from group B if it has a large amount of data to transmit and is operating under relatively good radio conditions as measured by the

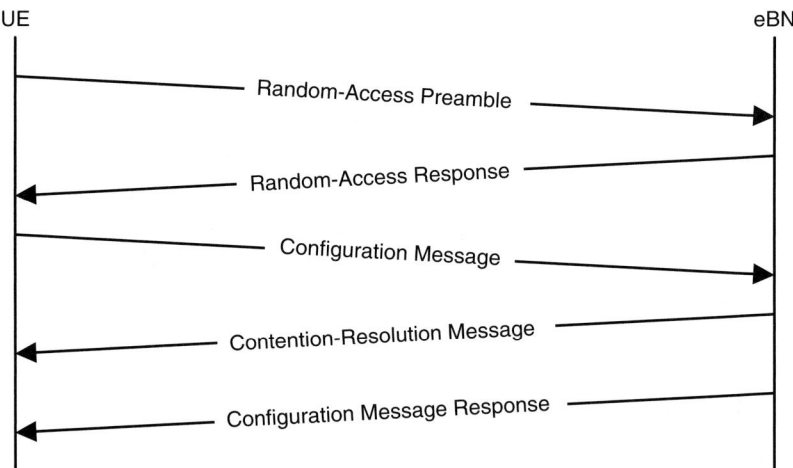

Figure 4.17. Random access procedure.

pathloss. Otherwise it will select a preamble from group A. The mobile then randomly selects a preamble from the group, determines the initial power setting using open-loop power control, and transmits the preamble.

Upon reception of the random-access preamble, the base station can send its response within a predefined timing window instead of at a specific time. This allows some scheduling flexibility and load balancing. This response is sent using a combination of L1/L2 control and downlink shared data channels. The L1/L2 control channel points to the location within the shared data channel where the actual random-access response is contained. Note that multiple responses may be multiplexed into the shared data channel. Each random-access response contains an uplink scheduling grant for data transmission, timing advance information, and an assignment of a temporary C-RNTI. Timing advance information is used by the mobile to time-synchronize its uplink transmission. Note that HARQ is not used for transmission of the random-access response due to the possibility of contention (i.e. more than one mobile selecting the same preamble sequence).

In the third step, the mobile transmits its message (e.g. an RRC connection request) in the uplink using its temporary C-RNTI. Included in this message is its mobile identifier and whether it has already been assigned a C-RNTI from a previous network access. This message is of dynamic size, and HARQ can be used to ensure that it is successfully received at the base station. The temporary C-RNTI will serve as its identity for contention-resolution purposes, which the base station would echo in the fourth message. This would serve as an early indication if a collision occurred during the previous transmission, and hence would allow the mobile to re-initiate the random-access procedure as soon as possible. Otherwise, the mobile will have to wait until the fifth message before contention is resolved. This may incur significant delay since the response to the RRC connection request has to come from the serving gateway.

4.7 Timing advance

In the uplink, transmission from users should be time-synchronized so that orthogonality is maintained. During initial access, UEs can obtain

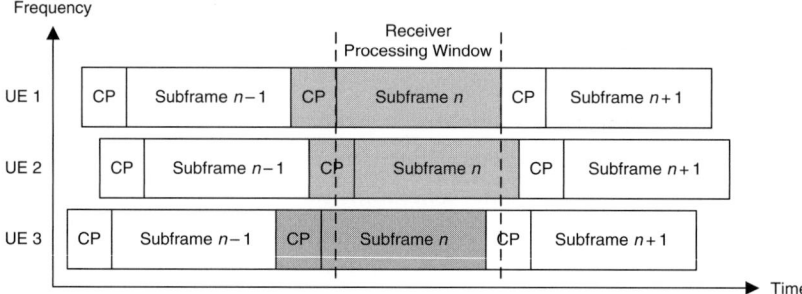

Figure 4.18. Uplink timing misalignment.

downlink timing and frequency synchronization on the basis of downlink synchronization signals. However, uplink transmission is not synchronized, with timing misalignment due to propagation delay relative to the eNB as well as channel profile characteristics. For example, a UE that is 500 m away from the cell has a timing misalignment of $2 \times 500/ (3 \times 10^8) = 3.33$ μs, which must be corrected. Note that the round-trip delay is used (hence the factor of 2) because the UE will lock onto the downlink synchronization signals from the eNB, which accounts for half of the round-trip delay. Once this initial access timing misalignment has been corrected, timing drift can occur due to either user movement or sudden changes in propagation channel. For example, a UE that is moving at 120 km/h will have a timing drift of around 1 μs per second. Additionally, sudden changes in propagation channel, for example turning a corner in a dense urban area, can also cause large changes in propagation channel leading to timing misalignment.

Without accurate timing synchronization, interference from different users within the same subframe as well as from surrounding subframes will degrade performance. To avoid such interference, signals from all uplink users must arrive in the demodulation window within a fraction of the cyclic prefix. The degree of performance degradation increases with timing error but, in general, a timing error of ±1 μs is deemed acceptable. An example of the timing misalignment among different users is shown in Figure 4.18. In this case, the receiver window is aligned with UE 1. The received signal from UE 2 is sufficiently time-aligned since it arrives

Figure 4.19. Timing measurement and required adjustment.

within the cyclic prefix. However, UE 3 is misaligned with the receiver window, and timing adjustment to delay the transmission from UE 3 is needed.

The eNB is responsible for maintaining uplink timing synchronization for the users. Thus, it must measure the uplink transmission timing and provide appropriate timing adjustment when necessary. When the UE is not synchronized, it must access the system using a random-access procedure, and timing information is obtained from the random-access preamble. Once the UE is synchronized and active in the system, the eNB can measure the timing alignment on the basis of any uplink signal transmission, including demodulation reference signals transmitted with PUCCH or PUSCH and SRSs, which may be configured to be transmitted periodically. On the basis of an uplink transmission, the eNB measures the channel profile against the receiver reference timing as shown in Figure 4.19. On the basis of this measurement, the eNB can issue an appropriate timing adjustment command to the UE.

In LTE, timing adjustment is performed using the timing-advance (TA) command, which is transmitted either as part of a random-access response or as a MAC control element. Timing adjustment is done in multiples of $0.52\,\mu s$. When the TA command is transmitted as part of the random-access response, an 11-bit TA command with possible values of $T_A = 0, 1, 2, \ldots, 1282$ is used. This provides timing adjustment from 0 to $667\,\mu s$ in steps of $0.52\,\mu s$, equivalent to a maximum supportable cell radius of $100\,km$. The UE then advances its uplink transmission by the given amount. This timing adjustment is used when the UE is completely unsynchronized. When the UE is already synchronized and only fine timing adjustment is necessary, the TA command is transmitted as part of a MAC control message. In this case, a 6-bit TA command with possible values of $T_A = 0$, 1, 2, ..., 63 is used and the UE adjusts its timing by $N_{TA,old} + (T_A - 31) \times 0.52 \, \mu s$. In this case, a negative value corresponds to delaying the uplink transmission.

4.8 Power control

In the uplink, power control refers to the setting of the uplink transmission power such that the spectral density per resource block is set to the appropriate level [9]–[11]. This is used to ensure that uplink transmission is received at the right power level for demodulation and also to limit the amount of interference with nearby cells. While the basic power-control concepts apply to all uplink channels, variations in their application are needed in order to deal with different characteristics and requirements.

4.8.1 Data channel

In the uplink, the UE transmit power is controlled by the eNB through controlling the power spectral density (PSD) per resource block. This power level is determined from several parameters given to the UE by the eNB. They include the number of resource blocks assigned, target power, pathloss and pathloss compensation factor, assigned modulation and coding level, and closed-loop adjustment. Some parameters, such as the target power and pathloss compensation factor, are cell-specific, meaning that the same values apply to all users in the same cell. Other parameters are UE-specific, meaning that different values may be assigned to different users. This is accomplished via the power-setting formula given by

$$P_{PUSCH} = \min(P_{CMAX}, 10 \log_{10}(M_{PUSCH}) + P_{O\_PUSCH} + \alpha \cdot PL + \Delta_{TF} + f(i)) \, [dBm]$$

where P_{CMAX} is the maximum configured UE transmit power and M_{PUSCH} is the number of PUSCH resource blocks used to scale the PSD per resource block. Several components are present in the power-setting formula – open-loop setting, closed-loop setting, and adjustment for the assigned transport format. The open-loop component includes the following parameters.

- P_{O_PUSCH} is the sum of a cell-specific target power broadcast by the eNB and a UE-specific parameter. This parameter can be thought of as the required SINR plus thermal noise and interference over thermal noise (IoT) term. UE-specific adjustment is used to correct any UE-specific errors (e.g. due to inaccurate power setting). Broadcast values of -126 to 24 dBm are allowed, with additional UE-specific adjustment of -8 to 7 dB.
- PL is the downlink pathloss estimated by the user. In LTE, the reference-signal transmit power is broadcast. This allows the pathloss to be calculated using the reference-signal received power (RSRP) measurement given in [12].
- α is a cell-specific fractional compensation factor, $\alpha \in \{0.0, 0.4, 0.5, 0.6, 0.7, 0.8, 0.9, 1.0\}$.

The parameters P_{O_PUSCH} and α are used to control the average SINR at the eNB. In LTE, fractional pathloss compensation is available via setting the parameter α. Fractional pathloss compensation, whereby the transmission power is set to compensate for only a portion of the pathloss, is beneficial for cell-edge users who generally have large pathloss and generate significant interference with neighboring cells. By limiting the transmit power via fractional pathloss compensation, the IoT seen in neighboring cells can be reduced significantly, providing improvements both in cell average and in cell-edge user throughput. Furthermore, since the UE transmit power consumption is correspondingly reduced, longer battery life can be achieved.

The long-term average SINR may be given as SINR $= P_{PUSCH} - 10 \log_{10}(M_{PUSCH}) - PL - IoT - N$, where N is the thermal noise. If just the open-loop power-control terms are considered, this SINR is given by SINR $= P_{O\_PUSCH} + (\alpha - 1) \cdot PL - IoT - N$. It can be seen that the SINR is therefore controlled through appropriate selection of P_{O_PUSCH} and α parameters. Note that the IoT is also dependent on P_{O_PUSCH} and α, and therefore changes in these parameters will affect the IoT as well.

In addition to open-loop power setting, closed-loop adjustment is available to compensate for fast fading, measurement errors, inaccuracies in mobile transmit power setting, varying interference level, etc. It also allows the desired SINR target for each UE to be individually controlled on the basis of, for example, quality-of-service requirements. The closed-loop component is denoted as $f(i)$.

- $f(i)$ is a UE-specific correction value that is applied to the power setting. Two modes are available to allow implementation-specific adjustments on the basis of the required convergence speed and power correction. Absolute adjustment is available in step sizes of $\{-4, -1, 1, 4\}$ dB, but can be done only when an uplink scheduling assignment is provided to the user. This mode provides fast power correction and avoids TPC propagation error. Accumulated adjustment can be done either via the uplink scheduling assignment or periodically via the TPC command in step sizes of $\{-1, 0, 1, 3\}$ dB. The value of $f(i)$ is obtained from the transmit power-control command given periodically or together with an uplink scheduling grant. The parameter $f(i)$ may be cumulative or absolute:

$$f(i) = \begin{cases} f(i-1) + \delta_{\text{PUSCH}}(i - K_{\text{PUSCH}}) \text{ for cumulative mode} \\ \delta_{\text{PUSCH}}(i - K_{\text{PUSCH}}) \text{ for absolute mode} \end{cases}$$

where $\delta_{\text{PUSCH}}(i - K_{\text{PUSCH}})$ is the transmit power-control command sent to the user in subframe $i - K_{\text{PUSCH}}$, where K_{PUSCH} denotes the corresponding delay.

Finally, the power level may be adjusted by the assigned transport format given by Δ_{TF}.

- Δ_{TF} is a UE-specific parameter that adjusts the power setting on the basis of the assigned modulation and coding rate in order to arrive at the appropriate SINR for the selected modulation and coding rate. The parameter Δ_{TF} is given by

$$\Delta_{\text{TF}} = \begin{cases} 10 \log_{10}((2^{\text{MPR} \cdot K_s} - 1)\beta_{\text{offset}}^{\text{PUSCH}}), deltaMCS\text{-}Enabled \text{ on} \\ 0, deltaMCS\text{-}Enabled \text{ off} \end{cases}$$

where *deltaMCS-Enabled* is a UE-specific parameter configured by the RRC. With *deltaMCS-Enabled* off, power adjustment on the basis of the MCS is turned off. Instead, MCS adjustment can be used for link adaptation. With *deltaMCS-Enabled* on, power is adjusted using the formula above, with $K_s = 1.25$. This adjustment is based on the observation that the spectral efficiency supported can be approximated by

$$\text{MPR} = \frac{1}{K_s} \log_2(1 + l \cdot \text{SINR})$$

Thus, the SINR required in order to support the assigned MPR is given by $\text{SINR} = (2^{\text{MPR} \cdot K_s} - 1)/l$. With the constant l incorporated into the $P_{\text{O\_PUSCH}}$ setting, the adjustment on the basis of the assigned transport format can be given by the formula above. The value of the parameter K_s depends on the receiver performance, with a typical value being $K_s = 1.25$.

4.8.2 Control channels

Power control of the PUCCH is similarly determined through controlling the PSD as given by the equation

$$P_{\text{PUCCH}} = \min(P_{\text{CMAX}}, P_{\text{O\_PUCCH}} + \text{PL} + h(n_{\text{CQI}}, n_{\text{HARQ}}) \\ + \Delta_{\text{F\_PUCCH}}(F) + g(i)) \,[\text{dBm}]$$

Fractional power compensation is not used since full compensation is required in order to meet the target SINR represented by the $P_{\text{O\_PUCCH}}$ parameter. However, because different PUCCH formats have different SINR requirements, the following adjustment parameters are needed.

- $h(n_{\text{CQI}}, n_{\text{HARQ}})$ is an adjustment parameter based on the number of CQI information bits being transmitted in the control channel. When only HARQ acknowledgment bits are transmitted, this parameter is set to 0.
- $\Delta_{\text{F\_PUCCH}}(F)$ is a power-adjustment parameter based on the transmitted PUCCH format. The adjustment is relative to the required power for PUCCH format 1a. For example, when PUCCH format 2 is transmitted, the possible adjustment values are $\{-2, 0, 1, 2\}$.

Similarly to power control for PUSCH, closed-loop adjustment is available. The closed-loop component is given by

- $g(i)$, which is a UE-specific adjustment value given by TPC commands. The value is accumulated from past TPC commands according to

$$g(i) = g(i-1) + \sum_{m=0}^{M} \delta_{\text{PUCCH}}(i - k_m)$$

where δ_{PUCCH} is the transmit power control previously sent to the user.

4.8.3 Random-access channel

The random-access channel is used to transmit a random-access preamble. The preamble received target power is given by the higher layer, and can be calculated using the initial received target power scaled to the appropriate preamble format. The transmission power is incremented in each subsequent transmission according to the power-ramping step size. Power setting of the PRACH is given by the equation

$$P_{\text{PUCCH}} = \min(P_{\text{CMAX}}, \text{PREAMBLE\_RECEIVED\_TARGET\_POWER} + \text{PL}) \, [\text{dBm}]$$

where the PRAMABLE_RECEIVED_TARGET_POWER is provided to the user via higher-layer signaling.

4.8.4 Sounding reference signal

Power setting of the SRS closely follows that of the PUSCH and is given by

$$P_{\text{SRS}} = \min(P_{\text{CMAX}}, P_{\text{SRS\_OFFSET}} + 10 \log_{10}(M_{\text{SRS}}) + P_{\text{O\_PUSCH}} + \alpha \\ \cdot \text{PL} + f(i)) \, [\text{dBm}]$$

where M_{SRS} is the SRS transmission bandwidth given in terms of the number of resource blocks. The term $P_{\text{SRS\_OFFSET}}$ is a cell-specific parameter. The parameters $P_{\text{O\_PUSCH}}$ and $f(i)$ are given in Section 4.8.1.

4.9 Interference coordination schemes

In the uplink, several inter-cell interference coordination (ICIC) schemes can be implemented. They include schemes based on fractional frequency reuse and schemes based on fractional power control. In general, ICIC schemes are used to improve cell-edge throughput. For fully loaded systems, this improvement in cell-edge user throughput is usually obtained at the expense of reduced overall cell throughput. However, for lightly loaded systems, both cell-edge user throughput and cell throughput can be improved. In addition, ICIC schemes can be used to improve system capacity for delay-sensitive traffic such as VoIP or video streaming.

Uplink ICIC schemes based on fractional frequency reuse are similar to the downlink schemes described in Section 3.9. In this case, eNBs reserve a fraction of the bandwidth for use by the cell-edge users. This can be done in a static or semi-static manner, but the main idea is that a different reuse pattern is assigned to each eNB such that the interference in this region is minimized. For example, in static ICIC with three eNBs belonging to the same physical site, each eNB is allowed to schedule users only on a predefined third of the resource blocks. In semi-static ICIC, the reuse pattern is allowed to vary on the basis of information exchanged among eNBs. In Rel-8, the uplink high-interference indicator (HII) information element is defined. The HII contains a bitmap of interference tolerance for all the resource blocks, where each eNB indicates whether the resource block can tolerate high interference or not. The HII information is shared among eNBs via the X2 interface. On the basis of information from the HII, each eNB can determine the optimal resource-block-reuse pattern for its own cell.

A second approach to uplink ICIC is fractional power-control adaptation, whereby the transmit power of the users within the cell is adaptively controlled to minimize the impact on the surrounding cells. In this case, different eNBs will adapt their own power-control parameters (usually the fractional power-control parameter α and the reference received power target $P_{O, \text{PUSCH}}$) on the basis of feedback from surrounding cells. In Rel-8, the uplink overload indicator (OI) information element is defined to assist in this power adaptation. The OI contains a bitmap of the observed interference level for each physical resource block. Three interference values are defined – high, medium, and low. Similarly to the HII, the OI information is shared among eNBs via the X2 interface. On the basis of information from the OI, each eNB can determine the optimal power-control parameters for its own cell. For example, each eNB can adapt its power-control parameters on the basis of how many high-interference OIs it received. The idea is to lower the interference level if neighboring cells report lots of high-level interference on their resource blocks. Performance results from [10] showed that this approach can provide a significant gain (~150%) in the cell-edge user throughput at the expense of a moderate loss (30%) of cell throughput in fully loaded systems.

4.10 Performance results

4.10.1 Link-level performance

Table 4.16 provides an illustrative example of the typical operating requirements as well as the SNR for various uplink channels [13]. For example, for PUCCH format 1 (1-bit ACK/NACK), a typical operating point is for the false-alarm and false-detection rate to be below 1%. For the data channel, the operating point is usually set at 10% BLER for the initial transmission, resulting in throughput of approximately 91% of the initial data rate when HARQ is considered. In Table 4.16, the required SNR per subcarrier is given for the typical urban (TU) propagation channel [14], which is widely used in analysis of cellular networks due to its accurate representation of an urban propagation environment [15]. Two receive antennas are present at the eNB, providing receiver diversity as well as combining gains.

An example of an uplink link budget based on the SNR requirements in Table 4.16 is shown in Table 4.17. The link budget provides the maximum allowable pathloss, which may then be mapped into cell-area coverage. The numbers provided are examples based on typical hardware performance and system conditions [16]. Note that the link performance can be improved with more advanced receiver algorithms such as turbo equalization [17]. From Table 4.17, it can be seen that coverage is balanced between the different control and random-access channels, and is generally limited by the required data rate at the cell edge. In this case, using VoIP service as the guideline, the maximum allowable pathloss is approximately 124 dB, which translates into a supportable cell size of

Table 4.16. *Uplink link-level channel performance (two receive antennas)*

Physical channel	Typical operating requirements	Typical SNR operating points (dB)
PRACH		
Preamble format 0	$P(FA) = 1\%, P(MD) = 1\%$	−9.1
Preamble format 2	$P(FA) = 1\%, P(MD) = 1\%$	−11.2
Preamble format 4	$P(FA) = 1\%, P(MD) = 1\%$	−1.8
PUCCH		
Format 1a	$P(FA) = 1\%$	−7.1
Format 2	1% BLER	−6.5
PUSCH		
VoIP, 12.2 AMR	10% BLER for first transmission	−4.2
	10% BLER for first transmission	−8.0 (Subframe bundling)
FTP, 5.2 Mbps	10% BLER for first transmission	2.5
FTP, 21.4 Mbps	10% BLER for first transmission	11.5
FTP, 36.7 Mbps	10% BLER for first transmission	19.0

approximately 0.78 km for a carrier frequency of 2 GHz using the 3GPP pathloss model [14]

$$L = 128.1 + 37.6 \times \log_{10}(r)$$

where r is given in kilometers. To provide a cell-edge data rate of 5.2 Mbps, however, the supportable cell size is reduced significantly, to 0.23 km. Similarly, a maximum cell radius of 0.08 km can be supported at a cell-edge data rate of 36.7 Mbps. As can be seen from the link budget, the supportable cell radius is limited by the desired uplink cell-edge data rate and the corresponding pathloss.

Table 4.17. *Uplink link-budget example*

Uplink channel	PRACH Format 0	PUCCH 1-bit ACK/NACK	PUCCH 4-bit CQI	PUCCH VoIP	PUSCH VoIP (Subframe bundling)	PUSCH FTP (5.2 Mbps)	PUSCH FTP (21.4 Mbps)	PUSCH FTP (36.7 Mbps)
UE EIRP power (dBm)	23	23	23	23	23	23	23	23
Transmit antenna gain (dBi)	−2.0	−2.0	−2.0	−2.0	−2.0	−2.0	−2.0	−2.0
EIRP (dBm)	21.0	21.0	21.0	21.0	21.0	21.0	21.0	21.0
Base-station sensitivity								
Antenna gain	17.0	17.0	17.0	17.0	17.0	17.0	17.0	17.0
Transmission line loss (dB)	3.0	3.0	3.0	3.0	3.0	3.0	3.0	3.0
BS noise figure (dB)	5.0	5.0	5.0	5.0	5.0	5.0	5.0	5.0
Thermal noise (kT) (dBm/Hz)	−174.0	−174.0	−174.0	−174.0	−174.0	−174.0	−174.0	−174.0
Bandwidth (kHz)	1080	180	180	360	360	9000	9000	9000
Required SNR (dB)	−9.1	−7.1	−6.5	−4.2	−8.0	2.5	11.5	19.0
Base-station sensitivity	−131.7	−137.5	−136.9	−131.6	−135.4	−110.9	−101.9	−94.4

Margins								
Lognormal fade margin	4.9	4.9	4.9	4.9	4.9	4.9	4.9	4.9
Interference margin	3.0	8.0	3.0	3.0	3.0	3.0	3.0	3.0
Penetration loss	18.0	18.0	18.0	18.0	18.0	18.0	18.0	18.0
MS body loss	2.0	2.0	2.0	2.0	2.0	2.0	2.0	2.0
Total system margin	27.9	32.9	27.9	27.9	27.9	27.9	27.9	27.9
Maximum allowable pathloss	124.8	125.6	130.0	124.7	128.5	104.0	95.0	87.5

4.10.2 System-level performance

In this section, the uplink system performance is provided for four distinct scenarios defined in [18] – an indoor hotspot, an urban micro-cell, an urban macro-cell, and a rural macro-cell. The micro-cell and macro-cell scenarios are traditional 19-cell, 57-sector system setups, while only two cells are present in the indoor hotspot scenario. The pathloss profiles of the four scenarios are shown in Figure 4.20. From the profile, it can be seen that the performance of the indoor hotspot will be limited by interference, whereas the urban micro-cell and rural macro-cell are noise-limited. For the urban macro-cell, however, the performance is both noise-limited and interference-limited.

Representative results for full-buffer traffic for the four scenarios are shown in Table 4.18. The setup assumes a system bandwidth of 10 MHz with two uncorrelated receive antennas at the eNB. The UE has only one transmit antenna. The results show the sector and cell-edge (denoted as the fifth-percentile user throughput) spectral efficiencies. So, for example, in an FDD 10-MHz urban micro-cell system, an average aggregate through-put of 12.3 Mbps can be expected per sector. At the same time, 5% of the users will experience throughput of 345 kbps or less. From the results, it can be seen that the system throughput follows an expected pattern with the respective scenarios. In general, the larger the cell size, the lower the sector and cell-edge throughput. In addition, for delay-insensitive traffic, the spectral efficiencies of FDD and TDD systems are similar. Additional uplink system-level performance results may be found in [19]–[20]. In comparison with HSUPA system-level results in [21], it is seen that LTE provides an improvement by a factor of 2–3 in system performance.

Table 4.19 summarizes the uplink VoIP capacity for the systems considered since LTE VoIP capacity is limited by the uplink [22]–[24]. In this case, performance is normalized to supportable VoIP users per MHz at 2% outage, where an outage is defined as 2% or more of the packets being lost or discarded. Thus, for example, when the capacity is given as 51 VoIP users per MHz, 510 VoIP users can be supported in a 10-MHz system. Results are provided both for FDD and for TDD configuration 1. It is seen here that results for normalized TDD systems in some scenarios are

Table 4.18. *Uplink system-level performance (full-buffer traffic)*

	FDD		TDD	
Scenario	Sector spectral efficiency (bps/Hz)	Cell-edge spectral efficiency (bps/Hz)	Sector spectral efficiency (bps/Hz)	Cell-edge spectral efficiency (bps/Hz)
Indoor hotspot	2.24	0.140	2.21	0.136
Urban micro-cell	1.32	0.035	1.21	0.030
Urban macro-cell	0.91	0.025	0.88	0.022
Rural macro-cell	0.87	0.021	0.86	0.019

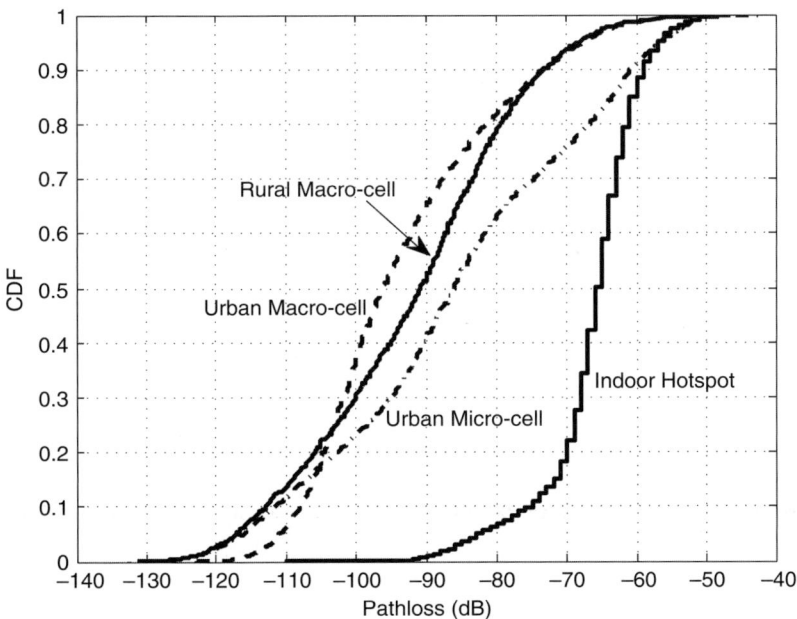

Figure 4.20. Pathloss of the scenarios considered.

Table 4.19. *Uplink system-level performance (VoIP)*

Scenario	Dynamic scheduling (users/MHz)		Semi-persistent scheduling (users/MHz)	
	FDD	TDD	FDD	TDD
Indoor hotspot	62	63	136	130
Urban micro-cell	51	52	78	70
Urban macro-cell	49	40	88	95
Rural macro-cell	47	40	97	105

significantly poorer than for FDD systems. This is because of the additional latency for retransmission that is incurred while waiting for an uplink subframe to become available.

Results for both dynamic and semi-persistent scheduling are shown in Table 4.19. Dynamic scheduling requires associated control information with every packet. Thus, control-channel limitation can restrict the number of VoIP users that may be scheduled in one subframe [25]–[26]. With semi-persistent scheduling, however, the VoIP capacity increases significantly. This is because semi-persistent scheduling can be used to assign resources in a periodic manner on the basis of the VoIP traffic pattern. Thus, for traffic types with regular data transmissions such as VoIP and real-time video, semi-persistent scheduling is very beneficial.

Table 4.20 illustrates the system performance as a function of the number of receive antennas at the eNB. Increasing the number of receive antennas provides diversity and combining gain. It can be seen that both sector and cell-edge performance can be improved significantly. For instance, for the urban micro-cell scenario, the sector throughput is increased by 41% and 81% when the number of receive antennas is four and eight, respectively.

Figure 4.21 illustrates the throughput CDF for different cell sizes and environments for urban micro-cell and rural macro-cell scenarios with two and four receive antennas. Significant performance improvement from additional receive antennas is apparent.

Table 4.20. *Uplink system-level performance (full-buffer traffic, FDD)*

Scenario	Two receive antennas		Four receive antennas		Eight receive antennas	
	Sector spectral efficiency (bps/Hz)	Cell-edge spectral efficiency (bps/Hz)	Sector spectral efficiency (bps/Hz)	Cell-edge spectral efficiency (bps/Hz)	Sector spectral efficiency (bps/Hz)	Cell-edge spectral efficiency (bps/Hz)
Indoor hotspot	2.24	0.140	2.79	0.175	3.12	0.212
Urban micro-cell	1.32	0.035	1.86	0.067	2.39	0.089
Urban macro-cell	0.91	0.025	1.37	0.053	1.90	0.086
Rural macro-cell	0.88	0.021	1.40	0.047	2.07	0.073

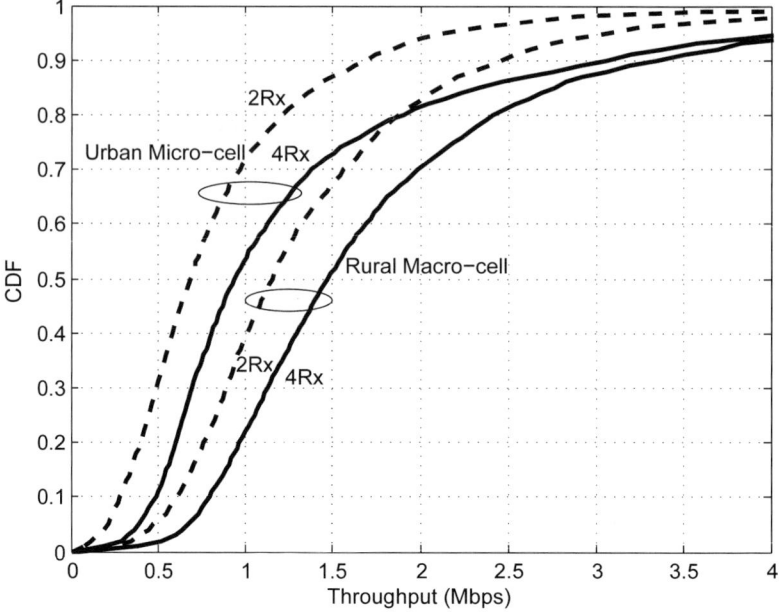

Figure 4.21. Throughput CDF for urban micro-cell and rural macro-cell (*n*Rx indicates the number of receive antennas).

From the link- and system-level results shown in this section, several observations may be made regarding uplink performance for LTE.

- LTE provides very high uplink data capacity. With four receive antennas at the eNB, a cell spectral efficiency of 1.4–2.8 bps/Hz can be achieved depending on the deployment scenario.
- LTE provides very high uplink voice capacity. With four receive antennas at the eNB, a VoIP capacity of 80–140 users per MHz can be achieved depending on the deployment scenario.
- The performance of FDD and TDD systems is similar for delay-non-sensitive traffic. For delay-sensitive traffic such as VoIP, however, the performance for TDD systems is worse due to retransmission latency incurred while waiting for an uplink sub-frame to become available.

References

[1] R1-070037, "DFT Restrictions and Impact on UL System Performance", Motorola, RAN1#47bis, Sorrento, Italy, January 2007.

[2] Classon, B., Baum, K., Nangia, V. *et al.*, "Overview of UMTS air-interface evolution," *IEEE 64th Vehicular Technology Conference*, September 2006.

[3] Susitaival, R., Meyer, M., "LTE coverage improvement by TTI bundling," *IEEE 69th Vehicular Technology Conference*, April 2009.

[4] Ghosh, A., Ratasuk, R., Xiao, W. *et al.*, "Uplink control channel design for 3GPP LTE," *IEEE 18th International Symposium on Personal, Indoor and Mobile Radio Communications*, September 2007.

[5] Rahman, M. I., Astely, D., "Link level investigation of ACK/NACK bundling for LTE TDD," *IEEE 69th Vehicular Technology Conference*, April 2009.

[6] Kishiyama, Y., Higuchi, K., Sawahashi, M., "Investigations on random access channel structure in Evolved UTRA uplink," *International Symposium on Wireless Communication Systems*, pp. 287–291, September 2006.

[7] Popovic, B. M., Mauritz, O., "Random access preambles for Evolved UTRA cellular system," *IEEE 9th International Symposium on Spread Spectrum Techniques and Applications*, pp. 488–492, August 2006.

[8] Ghosh, A., Ratasuk, R., Filipovich, I., Tan, J., Xiao, W., "Random access design for UMTS air-interface evolution," *IEEE 65th Vehicular Technology Conference*, pp. 1041–1045, April 2007.

[9] Simonsson, A., Furuskar, A., "Uplink power control in LTE – overview and performance," *IEEE 68th Vehicular Technology Conference*, September 2008.

[10] Xiao, W., Ratasuk, R., Ghosh, A. *et al.*, "Uplink power control, interference coordination and resource allocation for 3GPP E-UTRA," *IEEE Vehicular Technology Conference*, September 2006.

[11] Castellanos, C. U., Villa, D. L., Rosa, C. *et al.*, "Performance of uplink fractional power control in UTRAN LTE," *IEEE Vehicular Technology Conference*, pp. 2517–2521, May 2008.

[12] 3GPP TS 36.214, Physical layer – measurements, v8.6.0, March 2009.

[13] Dabbagh, A. D., Ratasuk, R., Ghosh, A., "On UMTS-LTE physical uplink shared and control channels," *IEEE 68th Vehicular Technology Conference*, September 2008.

[14] 3GPP TS 25.814, Physical layer aspects for evolved Universal Terrestrial Radio Access (UTRA), v7.1.0, September 2006.

[15] Asplund, H., Larsson, K., Okvist, P., "How typical is the "Typical Urban" channel model?," *IEEE Vehicular Technology Conference*, pp. 340–343, May 2008.

[16] Prasad, N., Shuangquan, W., Xiaodong, W., "Efficient receiver algorithms for DFT-spread OFDM systems," *IEEE Transactions on Wireless Communications*, vol. 8, no. 6, pp. 3216–3225, June 2009.

[17] Berardinelli, G., Priyanto, B. E., Sorensen, T. B., Mogensen, P., "Improving SC-FDMA performance by turbo equalization in UTRA LTE uplink," *IEEE Vehicular Technology Conference*, pp. 2557–2561, May 2008.

[18] ITU-R M.2135, Guidelines for evaluation of radio interface technologies for IMT-Advanced, 2008.

[19] Lunttila, T., Lindholm, J., Pajukoski, K., Tiirola, E., Toskala, A., "EUTRAN uplink performance," *International Symposium on Wireless Pervasive Computing*, February 2007.

[20] Wong, I. C., Oteri, O., McCoy, W., "Optimal resource allocation in uplink SC-FDMA systems," *IEEE Transactions on Wireless Communications*, vol. 8, no. 5, pp. 2161–2165, May 2009.

[21] Ghosh, A., Love, R., Whinnett, N. *et al.*, "Overview of enhanced uplink for 3GPP W-CDMA," *IEEE 59th Vehicular Technology Conference*, pp. 2261–2265, May 2004.

[22] Nory, R., Kuchibhotla, R., Love, R., Sun, Y., Xiao, W., "Uplink VoIP support for 3GPP EUTRA," *IEEE 65th Vehicular Technology Conference*, pp. 710–714, April 2007.

[23] Haiming, W., Dajie, J., Tuomaala, E., "Uplink capacity of VoIP on LTE system," *Asia–Pacific Conference on Communications*, pp. 397–400, October 2007.

[24] Jing, H., Haiming, W., "Principle and performance of TTI bundling for VoIP in LTE FDD mode," *IEEE Wireless Communications and Networking Conference*, April 2009.

[25] Puttonen, J., Puupponen, H.-H., Aho, K., Henttonen, T., Moisio, M., "Impact of control channel limitations on the LTE VoIP capacity," *Ninth International Conference on Networks (ICN)*, April 2010.

[26] Valkama, M., Anttila, L., Renfors, M., "Some radio implementation challenges in 3G-LTE context," *IEEE 8th Workshop on Signal Processing Advances in Wireless Communications*, June 2007.

[27] 3GPP TS 36.213, Physical layer procedures, v8,8,0, September 2009.

5 MIMO

5.1 Introduction

OFDM systems naturally benefit from the use of multi-antenna systems (MASs), which improves the capacity and coverage of the LTE system significantly. In the downlink, four different multi-antenna transmission techniques are supported – transmit diversity, closed-loop spatial multiplexing using precoding codebooks, open-loop spatial multiplexing, and user-specific reference-symbol-based beamforming [1]–[3]. Spatial multiplexing can be used to support single-user MIMO (SU-MIMO), whereby multiple data streams (or spatial layers in LTE terminology) are transmitted to the same user simultaneously in the same time–frequency resource, or multi-user MIMO (MU-MIMO), whereby multiple data streams (or layers) are transmitted to different users simultaneously using the same time–frequency resource. A significant gain in system capacity can be achieved with MIMO [4]. In the uplink, SU-MIMO is not possible since the UE can only transmit on one antenna. However, MU-MIMO can be supported in the uplink. In this chapter, a comprehensive description of various multi-antenna technologies for Rel-8 downlink and uplink is presented, together with details of their performance.

5.2 Downlink multi-antenna techniques

In LTE, each multi-antenna transmission technique is denoted by a transmission mode. There are seven transmission modes in downlink, corresponding to transmit diversity, open-loop spatial multiplexing, SU-MIMO closed-loop spatial multiplexing, MU-MIMO closed-loop spatial multiplexing, and UE-specific reference-symbol-based beamforming. The different downlink multi-antenna schemes for the data channel are summarized in Table 5.1.

Table 5.1. *Summary of various downlink transmission modes for the PDSCH*

MAS	Transmit mode	Precoding	Antenna		Preferred UE velocity	Antenna geometry	Reference symbols
			eNB	UE			
SISO	1	None	1	1, 2, 4	All	Uncorrelated	CRS
Open-loop transmit diversity	2	SFBC/FSTD	2, 4	1, 2, 4	>30 km/h	Uncorrelated	CRS
Open-loop spatial multiplexing	3	Large-delay CDD	2, 4	1, 2, 4	>30 km/h	Uncorrelated	CRS
Closed-loop spatial multiplexing	4, 6	SU-MIMO	2, 4	1, 2, 4	<30 km/h	Correlated/ uncorrelated	CRS
Multi-user MIMO	5	SU-MIMO codebook subset	2, 4	1, 2, 4	<30 km/h	Correlated	CRS
UE-specific RS-based beamforming	7	None	4, 8	1, 2, 4	<60 km/h	Correlated	UE-specific RS

There are generally three classes of multi-antenna transmission techniques that are used for the data channel, namely open-loop schemes using common reference symbols (CRSs) and a precoder, closed-loop schemes based on CRSs and a precoder, and a closed-loop scheme based on dedicated reference symbols (DRSs), which is mainly applicable for LTE-TDD. It may be noted that the CRSs and DRSs are used for channel estimation and demodulation of the downlink shared channel. The use of CRSs is limited to a maximum of four antennas in LTE since the overhead becomes excessive (>15%) if more than four CRSs are to be supported. All the Rel-8 transmission modes use CRSs, except for transmission mode 7, which uses UE-specific DRSs. However, DRSs will be used for advanced multiple-antenna schemes in LTE-A in order to further improve the performance of the downlink shared channel and also for overhead reduction. This is discussed in detail in Chapter 6. For the control and broadcast channels, the only technique supported is transmit diversity (using space frequency block code). The baseband signal generation for the data channel uses the following steps.

1. Scrambling of code bits in each codeword. A maximum of two codewords is supported.
2. Modulation of scrambled bits to generate complex-valued modulation symbols.
3. Mapping of the modulation symbols onto one or several transmission layers. A maximum of four layers is supported in LTE Rel-8.
4. Precoding of the modulation symbols on each layer for transmission on the antenna ports.
5. Mapping of the modulation symbols for each antenna port to resource elements.
6. Generation of a time-domain OFDM signal for each antenna port.

In LTE, different UEs in a cell are allowed to have different transmission modes, with the ability to do mode switching. As an example, a UE moving at slow speed starts with a default transmission mode 4 but, when it moves to the cell edge, the eNB scheduler has the ability to switch it to transmission mode 3 if the PMI feedback becomes unreliable. Next, the various downlink multi-antenna schemes are discussed in detail.

5.2.1 Transmission mode 2: transmit diversity

The SFBC operation, also called precoding for transmit diversity, is defined for two and four transmit antennas and used for rank-1 (single-stream) transmission. It is the default multi-antenna scheme for the common downlink control channels and is also used by the data channel. The SFBC scheme can be used with up to four antennas and uses common reference symbols. For the simple case of two antennas, the symbols are grouped into pairs as shown below and transmitted from the two antennas using two adjacent subcarriers S_1 and S_2 as given by the following equation:

$$\begin{bmatrix} A_{11} & A_{12} \\ A_{21} & A_{22} \end{bmatrix} = \frac{1}{\sqrt{2}} \begin{bmatrix} 1 & 1 \\ 1 & -1 \end{bmatrix} \times \begin{bmatrix} S_1 & -S_2^* \\ S_2 & S_1^* \end{bmatrix}$$

$$= \frac{1}{\sqrt{2}} \begin{bmatrix} S_1 + S_2 & -S_2^* + S_1^* \\ S_1 - S_2 & -S_2^* - S_1^* \end{bmatrix}$$

where A_{11} and A_{12} are the symbols transmitted from the first and second antennas on the first subcarrier and A_{21} and A_{22} are the symbols transmitted from the first and second antennas on the second subcarrier. The scheme is easily extended to four antennas. It may be noted that, with four-antenna SFBC, four common reference signals are used for demodulation of data symbols. Finally, eight-antenna SFBC is not possible in LTE Rel-8 since the number of CRSs is limited to four. In order to support transmit diversity with eight antennas, low-delay CSD with a delay of several samples between antenna pairs is used, followed by the LTE SFBC on the four antennas.

5.2.2 Transmission mode 3: precoder-based open-loop spatial multiplexing

The term open loop indicates that there is no feedback related to the precoding-matrix indicator (PMI) from the UE. Precoder-based open-loop spatial multiplexing with up to four layers is implemented using a linear precoding matrix, which is applied at the transmitter via large-delay cyclic delay diversity (CDD) operation. This feature is generally useful at relatively high vehicle speeds, at which PMI feedback is unreliable. For the case when two codewords S_1 and S_2 are spatially multiplexed using two transmit antennas the large-delay CDD operation is given by

$$
\begin{bmatrix} A_{11} \\ A_{12} \end{bmatrix} = \frac{1}{\sqrt{2}} \begin{bmatrix} 1 & 0 \\ 0 & 1 \end{bmatrix} \times \begin{bmatrix} 1 & 0 \\ 0 & e^{-j2\pi/2} \end{bmatrix} \times \frac{1}{\sqrt{2}} \begin{bmatrix} 1 & 1 \\ 1 & e^{-j2\pi/2} \end{bmatrix} \begin{bmatrix} S_1 \\ S_2 \end{bmatrix}
$$
$$
= \frac{1}{2} \begin{bmatrix} (S_1 + S_2) \\ (S_1 + S_2 e^{-j\pi}) \end{bmatrix}
$$

where A_{11} and A_{12} are the symbols transmitted from the first and second antennas on the first subcarrier. The first term in the equation is the precoding matrix for two-layer spatial multiplexing and the combination of the second and third terms is the large-delay cyclic diversity matrix. In the case of four transmit antennas, the number of layers can vary from one to four and eNB cyclically assigns different precoding matrices on the basis of the number of layers taken from a fixed-size codebook. Finally, it may be noted that the codeword-to-layer mapping is fixed for both open- and closed-loop spatial multiplexing. The scheme can easily be extended to more than two antennas, as defined in [5].

5.2.3 Transmission modes 4 and 6: closed-loop spatial multiplexing (single-user MIMO)

The conceptual diagram of a precoder-based SU-MIMO scheme is illustrated in Figure 5.1. In this scheme, the UE estimates the propagation channel from the common reference signals and computes the PMI from the codebook configured at the eNB and the UE. The codebook index and the associated rank indicator are then transmitted to the eNB using the uplink channel. The rank indicator indicates the number of layers and there is a fixed mapping between the number of layers and codewords.

The eNB then uses the corresponding codebook entry to precode the transmission. As an example, if the codebook index transmitted by the UE is 1 and the corresponding rank indicator is 2, the precoding for spatial multiplexing is defined as follows for two-codeword transmission:

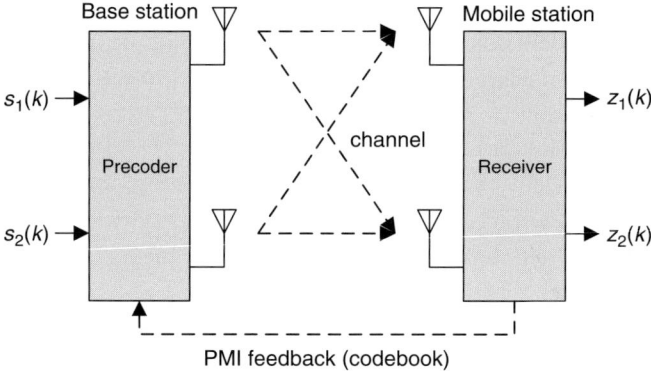

Figure 5.1. Illustration of the SU-MIMO scheme.

$$\begin{bmatrix} A_{11} \\ A_{12} \end{bmatrix} = \frac{1}{2} \begin{bmatrix} 1 & 1 \\ 1 & -1 \end{bmatrix} \begin{bmatrix} S_1 \\ S_2 \end{bmatrix} = \frac{1}{2} \begin{bmatrix} (S_1 + S_2) \\ (S_1 - S_2) \end{bmatrix}$$

where A_{11} and A_{12} are the symbols transmitted from the first and second antennas on the first subcarrier. However, the eNB can override the PMI feedback from the UE and replace it by its own PMI, which is indicated to the UE using the downlink control channel.

LTE supports 2×2, 4×2, and 4×4 downlink precoder-based SU-MIMO configurations with rank adaptation using up to four CRSs. The maximum numbers of layers and codewords supported are four and two, respectively. Transmission mode 6 is a special case of transmission mode 4 where the rank is constrained to 1, i.e. single-codeword transmission. The SU-MIMO scheme works very well at low to medium vehicle speeds using uncorrelated antennas, but is not useful at high vehicle speeds because of the unreliability of PMI feedback at high speeds. Interference, however, is a concern that can limit the multiplexing gain [6].

5.2.4 Transmission mode 5: multi-user MIMO

The MU-MIMO scheme in LTE Rel-8 is sub-optimal in nature and uses a subset of the SU-MIMO codebook. In this transmission mode, multiple UEs use the same time–frequency resource using rank-1 transmission.

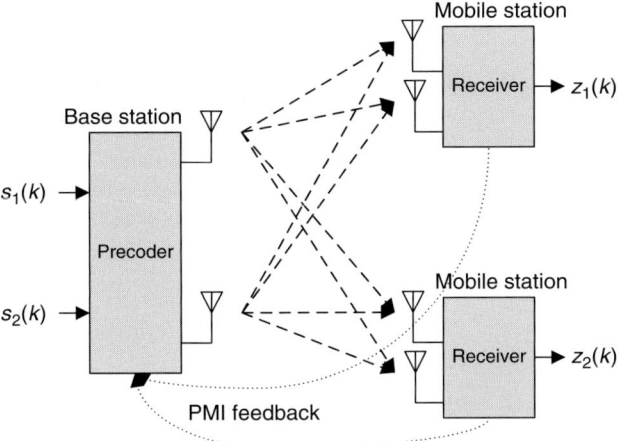

Figure 5.2. Illustration of the MU-MIMO scheme.

The CQI/PMI/RI feedback is the same as in the SU-MIMO scheme, and uses a 4-bit codebook-based feedback. The scheme is illustrated in Figure 5.2, where pairs of users feed back the codebook index and the eNB chooses the best user pair, predicts the MCS, and precodes the transmission using the codebook based on the transmitted PMI.

To achieve the maximum gain with MU-MIMO transmission, the eNB antennas need to be correlated. At the UE, there is no suppression of cross talk between the paired users. As will be shown later in this chapter, there is no performance advantage of this sub-optimal transmission mode compared with the SU-MIMO scheme (transmission modes 4 and 6). An advanced MU-MIMO scheme will be available in Rel-9 and Rel-10 of LTE, which will be described in Chapter 6.

5.2.5 Transmission mode 7: UE-specific reference-symbol-based beamforming

This mode is generally useful for LTE TDD (especially in LTE Rel-8) if the number of transmit antennas is greater than or equal to four. The eNB can semi-statically configure a UE to use the UE-specific reference signal,

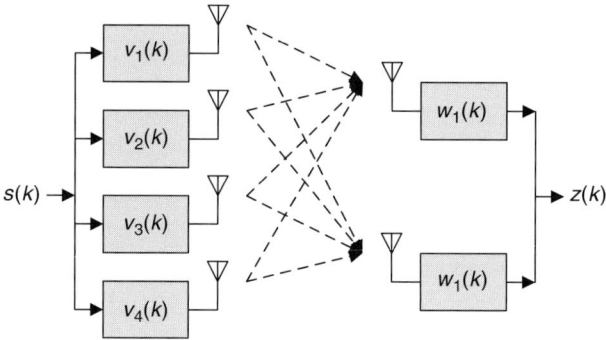

Figure 5.3. Baseband representation of transmission mode 7.

which acts as a phase reference for data demodulation of a single code-word (i.e. single-layer transmission) at the UE. At the eNB transmitter, a single set of transmit weights is computed, and these weights are applied to each subcarrier within a desired band. A covariance matrix of the channel is first computed over the band of interest, and the transmit antenna weights (denoted by vector $v_i(k)$ in Figure 5.3) are computed by taking the largest eigenvector of this covariance matrix, and applying it to all the data and UE-specific dedicated reference-symbol subcarriers within the band. Since the same weight is applied to the data and to the reference signal at the eNB, the UE is not required to have knowledge of the transmit weights for decoding the data. In TDD mode, the reciprocity principle allows the transmit weights to be computed on the basis of the uplink sounding reference signal. The baseband representation of transmission mode 7 is shown in Figure 5.3. The number of UE-specific reference symbols per PRB is 12 and their locations are shown in Figure 3.12.

In this mode, antenna-array calibration is required by the transmit antenna array in order to account for the variations in the gain and phase responses of the transmit and receive hardware across the multiple antennas. A separate calibration antenna circuitry is used at the eNB, which uses the following basic procedure for calibration.

1. The uplink and downlink channel responses between the calibrating antenna and each antenna of the base array are measured.

Table 5.2. *System simulation scenarios*

Simulation case	Inter-site distance (m)	Penetration loss (dB)	Speed (km/h)
3GPP case 1	500	20	3
3GPP case 3	1732	20	3
Urban micro-cell	200	20	3
Urban macro-cell	500	20	30
Rural macro-cell	1732	10	120
Indoor hotspot	–	0	–

2. Dividing the downlink channel response by the uplink channel response cancels out the reciprocal portions of the channel response, leaving the gain and phase differences between the transmit and receive hardware.
3. These gain and phase differences are then incorporated into the downlink closed-loop transmission strategy to account for the non-reciprocity of the antenna-array hardware.

5.2.6 System performance of LTE Rel-8 multiple-antenna schemes

In this section, aspects of the performance of various multiple-antenna schemes are compared via system simulations. The simulation cases are summarized in Table 5.2.

The performance details of the various MIMO modes are first compared for 3GPP case 1 for correlated and uncorrelated antennas with respect to sector and edge throughputs at various vehicle speeds as shown in Tables 5.3 and 5.4. Both 2×2 and 4×2 MIMO configurations are considered in the performance comparison for full-buffer traffic.

From Tables 5.3 and 5.4, the following observations are made.

1. The performance of 4×2 SU-MIMO (transmission mode 4) is approximately 25% better than that of 2×2 SU-MIMO (transmission mode 4), while the performance of 8×2 single-layer beamforming (transmission mode 7) is 25% better than that of 4×2 SU-MIMO.

Table 5.3. *Throughput comparison for downlink MAS with correlated*
0.5λ-spacing antennas

MIMO configuration	Sector throughput (kbps)			Cell-edge throughput (kbps)		
	3	30	120	3	30	120
EBF (four transmitters)	19 908	17 136	16 222	831.2	555.9	488.7
EBF (eight transmitters)	22 438	20 100	19 285	1073.8	681.7	600.5
SU-MIMO (rank-1 only)	19 816	17 017	16 066	823.4	573.2	498.5
SU-MIMO (rank-1 and rank-2)	21 139	17 339	16 110	802.9	571.2	499.9
OL-MIMO (four transmitters)	14 880	11 475	10 389	331.4	239.7	193.6
OL-MIMO (two transmitters)	14 254	10 823	9 750	328.0	226.6	179.9
SU-MIMO (two transmitters)	17 255	13 418	12 308	563.4	356.3	300.6

2. With a correlated antenna array, closed-loop MIMO outperforms open-loop MIMO in terms of both sector throughput and cell-edge throughput at all speeds. This is due to the fact that, with a correlated antenna array, there is very little transmit diversity or spatial multiplexing gain to be exploited by open-loop MIMO.

3. With UE-specific reference-signal-based beamforming (transmission mode 7), there is approximately 10% improvement in sector throughput and 25% improvement in edge throughput as the number of antennas at the eNB increases from four to eight. Finally, the degradation seen at high speeds is mainly from outdated CQI information, which affects both closed-loop and open-loop MIMO.

Table 5.4. *Throughput comparison for downlink MAS with large-spacing antennas*

MIMO configuration	Sector throughput (kbps)			Cell-edge throughput (kbps)		
	3	30	120	3	30	120
EBF (four transmitters)	15 037	10 841	9 578	457.4	231.4	219.1
EBF (eight transmitters)	16 579	12 069	10 515	587.5	331.4	255.3
SU-MIMO (rank-1 only)	15 499	10 079	8 089	485.0	283.1	207.3
SU-MIMO (rank-1 and rank-2)	18 968	11 115	8 904	433.7	263.6	189.1
OL-MIMO (four transmitters)	16 380	13 072	11 963	173.6	187.5	173.6
OL-MIMO (two transmitters)	17 233	12 340	11 258	228.8	226.7	216.1
SU-MIMO (two transmitters)	17 157	10 779	9 194	390.7	247.7	193.2

4. For uncorrelated antennas, the open-loop MIMO sector throughput is better than that for closed-loop MIMO except for the 4×2 configuration at low speeds.
5. With uncorrelated antennas, SU-MIMO outperforms open-loop MIMO for cell-edge throughput due to beamforming gain, though the difference disappears with increasing speed. In addition, for $n \geq 4$ transmit antennas, the sector throughput is always lower than for the correlated-antenna case.

Next the details of the performance of transmission modes 4 and 7 are compared for ITU environments for 2×2, 4×2, and 8×2 (transmission mode 7 only) cross-pole and uniform linear array (ULA) antenna configurations, respectively, and TDD configuration 1. The results are summarized in Tables 5.5–5.7.

Table 5.5. *2 × 2 SU-MIMO with cross-poles: TDD configuration 1*

Scenario	Cell spectral efficiency (bps/Hz)	Cell-edge user spectral efficiency (bps/Hz)
Indoor hotspot	4.42	0.158
Urban micro-cell	2.09	0.062
Urban macro-cell	1.09	0.020
Rural macro-cell	1.38	0.029

Table 5.6. *4 × 2 SU-MIMO using a pair of cross-poles with 0.5λ spacing: TDD configuration 1*

	Antenna configuration			
	Pair of cross-poles with 0.5λ spacing		Uniform linear array with 0.5λ spacing	
Scenario	Cell SE (bps/Hz)	Cell-edge user SE (bps/Hz)	Cell SE (bps/Hz)	Cell-edge user SE (bps/Hz)
Urban micro-cell	2.47	0.073	2.20	0.072
Urban macro-cell	1.48	0.030	1.54	0.036
Rural macro-cell	1.78	0.042	1.91	0.049

SE, spectral efficiency.

The following observations are made from Tables 5.5–5.7.

- The sector throughput performance of 2 × 2 and 4 × 2 transmission mode 4 (SU-MIMO) is superior to that of transmission mode 7 with any antenna configuration. This is due to the fact that transmission mode 7 supports only one stream, as opposed to two streams in the case of transmission mode 4.

Table 5.7. *UE-specific reference-signal beamforming performance (transmission mode 7) for various antenna configurations for the urban micro-cell scenario: TDD configuration 1*

Antenna configuration	Cell spectral efficiency (bps/Hz)	Cell-edge user spectral efficiency (bps/Hz)
2 × 2 Cross-pole	1.63	0.055
4 × 2 ULA	2.00	0.083
4 × 2 Cross-pole	1.90	0.072
8 × 2 Cross-pole	2.19	0.098

Table 5.8. *Performance at 0.5λ and 10λ antenna spacing (two transmit antennas, MIMO), FDD, case 1*

MIMO configuration	Antenna spacing 0.5λ		Antenna spacing 10λ	
	Cell SE (bps/Hz)	Cell-edge user SE (bps/Hz)	Cell SE (bps/Hz)	Cell-edge user SE (bps/Hz)
Rank 1 BF	1.94	0.074	1.60	0.051
SU-MIMO with MMSE	1.96	0.075	1.83	0.052
SU-MIMO with MMSE + SIC	2.01	0.075	1.94	0.048
MU-MIMO	2.00	0.074	1.60	0.049

SE, spectral efficiency.

• A cross-pole antenna is superior to a ULA for transmission mode 4, whereas in the case of beamforming (transmission mode 7) a ULA is superior.

Finally, the performance of closed-loop spatial multiplexing schemes and MU-MIMO (transmission mode 5) is compared for cross-pole and ULA cases with two and four transmit antennas. The results are summarized in Table 5.8. From the results shown, it can be seen that MU-MIMO works well only with correlated antennas, while SU-MIMO with rank adaptation works well both for uncorrelated and for correlated antennas. For rank-1 beamforming, correlated antennas provide better performance than do uncorrelated antennas. Additional performance analysis of downlink MIMO in LTE may be found in [7]–[17].

5.3 Uplink multi-antenna techniques

In LTE Rel-8, although multiple receive antennas are present at the UEs, only one transmit antenna is available. This is because only one power amplifier and transmitter chain is used in order to minimize cost and simplify the hardware design. Antenna selection, whereby the UE selects which of the antennas it will use for transmission, however, is enabled in the standards. In this mode, either the eNB can inform the UE of the preferred antenna for transmission, or the UE can autonomously select the transmit antenna. In either case, the UE will not transmit on more than one antenna at the same time. As a result, multi-antenna transmission techniques such as transmit diversity and precoding are not applicable to the uplink LTE. In LTE-A, however, multi-antenna transmission by the UEs will be supported.

At the eNB, however, multiple receive antennas (generally two, four, or eight) are present, and can be used to improve performance via receive diversity or multi-user spatial multiplexing. Receive diversity can improve performance by providing several independent radio-propagation channels (where a channel is from the transmit antenna to one of the receive antennas) for the transmitted signal. For instance, when one channel is experiencing severe fading, another channel may be in good condition. As a result, there is less fluctuation in the received signal and quality is improved. Performance improvement with multiple receive antennas is discussed in Section 3.8.2. Multi-user spatial multiplexing, on the other hand, is used to multiplex different users onto the same time and frequency resources in the spatial domain. This technique is commonly referred to as spatial division multiple access (SDMA). In LTE, however, the spatial multiplexing mode is also referred to as multi-user multiple-input multiple-output (MU-MIMO). MU-MIMO can also be viewed as a virtual MIMO system where, instead of multiple data streams being transmitted by a single user, the multiple data streams are coming from

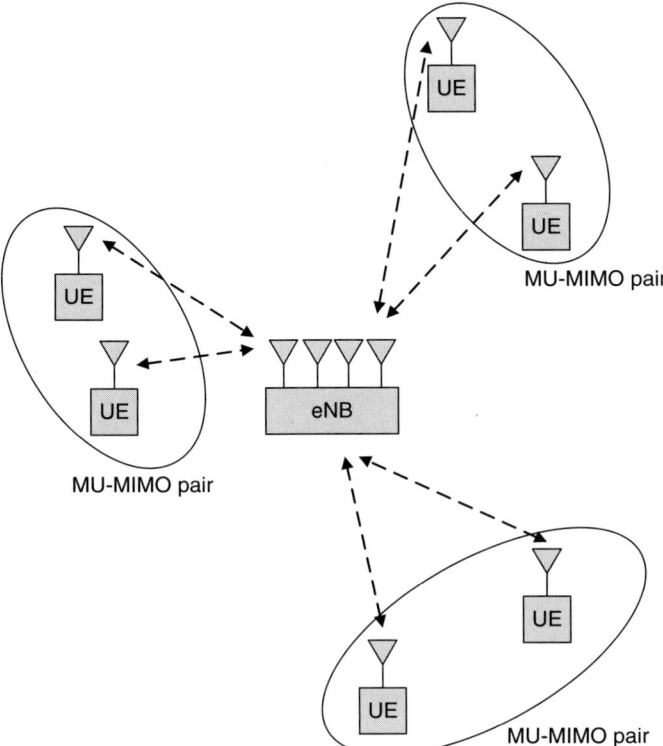

Figure 5.4. Uplink MU-MIMO in LTE.

different users [18]. For instance, two users may be scheduled to transmit on the same set of resource blocks in the same subframe. Sophisticated signal-processing techniques are then used at the eNB to separate out the signals and decode the two data packets from the two users [19]. As a result, spatial multiplexing can increase system throughput beyond what can be achieved with receive diversity alone.

A diagram illustrating uplink MU-MIMO is shown in Figure 5.4. In this example, two users are paired together to form a MU-MIMO pair. Each user within the MU-MIMO pair is then scheduled for uplink transmission to the eNB in the same subframe using the same resource blocks. Different MU-MIMO pairs can be scheduled simultaneously, depending on the number of resource blocks available. The pairing is also done dynamically and can change from subframe to subframe. Note that, although in the example two users are paired together, reference-signal design in LTE allows up to eight users to share the same resources. In theory, the maximum possible number of users sharing the same resources depends on the number of receive antennas at the eNB. For instance, with four receive antennas, up to four users can be multiplexed together. In practice, though, the multiplexing capability is much lower due to implementation constraints such as limited feedback information and eNB processing capabilities.

Uplink MU-MIMO is implemented in LTE standards in a manner that is transparent to the users. Each user is independently controlled by the eNB through scheduling assignment using DCI format 0 as described in Section 4.3.1 and does not know that MU-MIMO is being used. The transport block size, modulation, and coding rates are controlled independently, allowing separate link adaptation for each scheduled user. Furthermore, the resources assigned among MU-MIMO user pairs need not be the same because resource blocks can overlap among the different multiplexed users. When the multiplexed users share the same resource blocks, the reference signals are orthogonal, thus providing good channel-estimation performance. To ensure that the reference signals from the multiplexed users can be separated by the eNB, different cyclic shifts of the reference signal can be assigned to different users. With partially over-lapping assignment, however, the reference signals are no longer orthogo-nal. However, since the reference signals in LTE have low cross-correlation, sufficiently good channel-estimation performance can still be expected.

At the eNB, various user-pairing techniques may be used. They include, for example, random pairing, capacity pairing, orthogonal pair-ing, cross-correlation paring, bit-error-rate paring, and error-variance pairing [20]. In general, the more uncorrelated the transmissions from the paired users, the better the performance. Cross-correlation pairing, for instance, requires knowledge of the channel cross-correlation in order to pair users with orthogonal channels together. This minimizes perform-ance degradation arising from interference among the multiplexed users. This method, however, requires estimates of the cross-correlation, and also a sufficient number of users must be available for scheduling so that a good match can be found. Similarly, orthogonal pairing selects users with orthogonal radio channels in order to eliminate interference. This requires channel knowledge that can be estimated using the sounding reference signals, which will incur overhead. In practice, user pairing is a function of the scheduler and thus is performed at the MAC layer. As a result, channel state information may be stale by the time the scheduler is ready to do the pairing. Random pairing, on the other hand, selects users for multiplexing randomly and does not require knowledge of the user channel statistics. Random pairing is simplest to implement and can provide a reasonable performance gain.

In conjunction with user pairing, a strategy for power allocation or sharing is crucial in uplink MU-MIMO. This is because the users in general will interfere with each other, and interference management via power allocation is important for throughput improvement. In general, interference is controlled by limiting the total power output from all users that are multiplexed onto the same resource blocks. This requires that these users are allocated power subject to the sum not exceeding a limit. Without proper power allocation, MU-MIMO can lead to a significant increase in interference, negating or even reversing the benefits of MU-MIMO. Several power-allocation techniques are possible – uniform, maximum capacity, and minimum sum error rate [21]–[22]. For instance, with uniform power allocation, power is shared equally among the multiplexed users subject to the sum of all powers being smaller than a predetermined limit. This simple strategy does not take into account the channel qualities of different users or whether one user can more efficiently use the power than can another. A different strategy is the maximum-capacity approach, whereby power is allocated in a manner that will maximize the sum throughput (given by the sum of all the individual data rates). In this case, power is divided unequally, with most of the power being given to users in good channel conditions.

At the eNB, data transmissions from the multiplexed users are treated as independent spatially multiplexed streams. The eNB can use a conventional receiver such as the minimum mean-squared error (MMSE) receiver to try to decode each data stream separately, or it can employ more advanced receivers such as an interference rejection combining

Table 5.9. *Uplink system-level performance (full-buffer traffic, FDD, four receive antennas)*

	1 × 4		1 × 4 + MU-MIMO	
Scenario	Sector spectral efficiency (bps/Hz)	Cell-edge spectral efficiency (bps/Hz)	Sector spectral efficiency (bps/Hz)	Cell-edge spectral efficiency (bps/Hz)
Indoor hotspot	2.8	0.17	3.8	0.25
Urban micro-cell	1.9	0.07	2.3	0.08
Urban macro-cell	1.4	0.05	1.6	0.06
Rural macro-cell	1.4	0.05	1.6	0.05

(IRC) receiver or a successive interference canceller (SIC) receiver. The SIC receiver decodes different data streams successively while attempting to cancel out interference using information that had been decoded during previous attempts. Even more advanced receivers such as the maximum likelihood detector (MLD) are also possible, and implementation is limited solely by the processing capability of the eNB.

Uplink system performance is shown in Table 5.9 for four distinct scenarios (indoor hotspot, urban micro-cell, urban macro-cell, and rural macro-cell) as defined in Section 3.8.2. Up to two users may be multiplexed on the same resource blocks and power is shared equally between the multiplexed MU-MIMO users. In this case, a simple random user pairing is used, and the eNB uses a SIC receiver capable of performing interference cancellation. The users are power-controlled to limit the IoT to approximately 10 dB or less, depending on the deployment scenario. A typical cellular layout with 57 cells (19 sites with 3 cells per site) is used, with 10 active users per cell doing full-buffer transfer. The system bandwidth is 10 MHz and a frequency-reuse factor of 1 (full frequency reuse) is employed. Four receive antennas with antenna separation of 10 wavelengths (corresponding to uncorrelated antennas) are used at the eNB. Both sector and cell-edge spectral efficiencies are shown in Table 5.9.

From the results, it can be seen that MU-MIMO can increase both sector and cell-edge performance, with the amount of improvement being dependent on the deployment scenario. For instance, in an indoor hotspot, the sector throughput increases by 38% while the cell-edge throughput increases by 41%. In the urban micro-cell deployment scenario, the sector throughput increases by 24% while the cell-edge throughput increases by 19%. Also, even though MU-MIMO does not improve the signal strength of the users, there is a performance improvement also for users at the cell edge because more transmission opportunities are available. Furthermore, from the results shown in Table 5.9, it can be seen that the performance improvement for MU-MIMO is generally greater for scenarios in which users experience good channel conditions. This is because greater improvement can be obtained for users in good channel conditions due to the lower transmission power (and therefore less interference generated) and better channel-estimation performance. Note that additional performance gain may possibly be achieved by employing more sophisticated user-pairing and power-allocation strategies. In addition, MU-MIMO performance improves with the number of receive antennas at the eNB since the additional degrees of freedom minimize interference among the multiplexed users. Thus, performance can be significantly improved if eight receive antennas are used.

References

[1] Ghosh, A., Xiao, W., Ratasuk, R., Rottinghaus, A., Classon, B., "Multi-antenna system design for 3GPP LTE," *IEEE International Symposium on Wireless Communication Systems*, pp. 478–482, October 2008.

[2] Ghosh, A., Ratasuk, R., "Multi-antenna systems for LTE eNodeB," *IEEE 70th Vehicular Technology Conference*, September 2009.

[3] Qinghua, L., Guangjie, L., Wookbong, L. *et al.*, "MIMO techniques in WiMAX and LTE: a feature overview," *IEEE Communications Magazine*, vol. 48, no. 5, pp. 86–92, May 2010.

[4] Zihuai, L., Vucetic, B., Jian, M., "Ergodic capacity of LTE downlink multiuser MIMO systems," *IEEE International Conference on Communications, 2008*, pp. 3345–3349, May 2008.

[5] 3GPP TS 36.211, Physical Channels and Modulation, v8.7.0, May 2009.

[6] Andrews, J. G., Wan, C., Heath, R. W., "Overcoming interference in spatial multiplexing MIMO cellular networks," *IEEE Wireless Communications*, vol. 14, no. 6, pp. 95–104, December 2007.

[7] Ketonen, J., Juntti, M., Cavallaro, J. R., "Performance–complexity comparison of receivers for a LTE MIMO–OFDM system," *IEEE Transactions on Signal Processing*, vol. 58, no. 6, pp. 3360–3372, June 2010.

[8] Na, W., Pokhariyal, A., Sorensen, T. B., Kolding, T., Mogensen, P., "Performance of spatial division multiplexing MIMO with frequency domain packet scheduling: from theory to practice," *IEEE Journal on Selected Areas in Communications*, vol. 26, no. 6, pp. 890–900, August 2008.

[9] Hojin, K., Jianjun, L., Yongxing, Z., Kim, J. S., "On the performance of limited feedback single-/multi-user MIMO in 3GPP LTE systems," *3rd International Symposium on Wireless Communication Systems*, pp. 684–688, September 2006.

[10] Virtej, E., Kuusela, M., Tuomaala, E., "System performance of single-user MIMO in LTE downlink," *IEEE 19th International Symposium on Personal, Indoor and Mobile Radio Communications*, September 2008.

[11] Zihuai, L., Pei, X., Vucetic, B., "SINR distribution for LTE downlink multiuser MIMO systems," *IEEE International Conference on Acoustics, Speech and Signal Processing, 2009*, pp. 2833–2836, April 2009.

[12] Kian, C. B., Doufexi, A., Armour, S., "On the performance of SU-MIMO and MU-MIMO in 3GPP LTE downlink," *IEEE 20th International Symposium on Personal, Indoor and Mobile Radio Communications*, pp. 1482–1486, September 2009.

[13] Xiao, P., Lin, Z., Cowan, C., "Analysis of channel capacity for LTE downlink multiuser MIMO systems," *IEEE 72nd Vehicular Technology Conference*, September 2010.

[14] Wang, J., Wu, M., Zheng, F., "The codebook design for MIMO precoding systems in LTE and LTE-A," *6th International Conference on Wireless Communications Networking and Mobile Computing*, September 2010.

[15] Werner, K., Furuskog, J., Riback, M., Hagerman, B., "Antenna configurations for 4 × 4 MIMO in LTE – field measurements," *IEEE 71st Vehicular Technology Conference*, May 2010.

[16] Simonsson, A., Qian, Y., Ostergaard, J., "LTE downlink 2×2 MIMO with realistic CSI: overview and performance evaluation," *IEEE Wireless Communications and Networking Conference*, April 2010.

[17] Talukdar, A., Mondal, B., Cudak, M., Ghosh, A., Fan W., "Streaming video capacity comparisons of multi-antenna LTE systems," *IEEE Vehicular Technology Conference*, May 2010.

[18] Hanguang, W., Haustein, T., "Sum rate analysis of SDMA transmission in single carrier FDMA system," *11th IEEE Singapore International Conference on Communication Systems*, pp. 846–850, November 2008.

[19] Meili, Z., Bin, J., Ting, L., Wen, Z., Xiqi, G., "DCT-based channel estimation techniques for LTE uplink," *IEEE 20th International Symposium on Personal, Indoor and Mobile Radio Communications*, pp. 1034–1038, September 2009.

[20] Ruder, M. A., Dang, U. L., Gerstacker, W. H., "User pairing for multiuser SC-FDMA transmission over virtual MIMO ISI channels," *IEEE Global Communications Conference*, 2009.

[21] Haipeng, L., Xiaoqiang, L., "System level study of LTE uplink employing SC-FDMA and virtual MU-MIMO," *IEEE International Conference on Communications Technology and Applications*, pp. 152–156, October 2009.

[22] Hanguang, W., Haustein, T., "Sum rate analysis of SDMA transmission in single carrier FDMA system," *11th IEEE Singapore International Conference on Communication Systems*, pp. 846–850, November 2008.

6 LTE-Advanced

6.1 Introduction

Rel-8 LTE delivers improved system capacity and coverage, improved user experience through higher data rates, reduced-latency deployment, and reduced operating costs, and seamless integration with existing systems. Further enhanced requirements, however, were approved in 2008 to allow LTE to be approved as a radio technology for International Mobile Telecommunications-Advanced (IMT-Advanced). IMT-Advanced requirements are defined by the International Telecommunication Union, which is an organization that provides globally accepted standards for telecommunications. This further advancement for LTE is known as LTE-Advanced (LTE-A). The LTE-A requirements are shown in Table 6.1 and focus mainly on improvements in system performance and latency reduction. From Table 6.1, it can be seen that the target cell and user spectral efficiencies have increased significantly. Peak data rates of 1 Gbps in the downlink and 500 Mbps in the uplink must be supported. Target latencies have been significantly reduced as well. In addition to advancements in system performance, deployment and operating-cost-related goals were also introduced. They include support for cost-efficient multi-vendor deployment, power efficiency, efficient backhaul, open interfaces, and minimized maintenance tasks. A comprehensive list of LTE-A requirements can be found in [1].

To achieve these LTE-A requirements related to system performance, numerous physical-layer enhancements have been introduced in LTE-A [2]–[3]. They include carrier aggregation, enhanced downlink spatial multiplexing, uplink spatial multiplexing, and support for heterogeneous networks. Carrier aggregation allows multiple carriers to be aggregated to provide bandwidth extension up to 100 MHz. This provides a significant increase in the peak data rates, allows efficient interference management,

Table 6.1. *LTE-A requirements*

Feature	Requirements
Peak data rate	Downlink – 1 Gbps
	Uplink – 500 Mbps
Peak spectral efficiency	Downlink – 30 bps/Hz (8 × 8)
	Uplink – 15 bps/Hz (4 × 4)
Average cell spectrum efficiency	Downlink – 3.7 bps/Hz (4 × 4)
	Uplink – 2.0 bps/Hz (2 × 4)
Cell-edge user spectral efficiency	Downlink – 0.12 bps/Hz (4 × 4)
	Uplink – 0.07 bps/Hz (2 × 4)
C-plane latency	50 ms from camped to active state
	10 ms from dormant to active state
U-plane latency	Reduced compared with Rel-8

and supports heterogeneous deployment. Enhanced downlink spatial multiplexing extends the number of simultaneous data streams from four to eight, and allows coordinated data transmissions among eNBs. Uplink spatial multiplexing introduces SU-MIMO in the uplink and will allow four simultaneous data streams to be transmitted from a user. Heterogeneous networks consist of a traditional macro-cell-based network augmented with various types of low-power network nodes that address the capacity and coverage challenges resulting from the growth of data services. In this chapter, an overview of each of these LTE-A features is provided.

6.2 Carrier aggregation

Carrier aggregation is a feature in LTE-A to enable bandwidth extension to support deployment bandwidths of up to 100 MHz. This is done by aggregating several carriers to provide a larger system bandwidth [4]–[5]. It will allow LTE-A target peak data rates in excess of 1 Gbps in the downlink and 500 Mbps in the uplink to be achieved [1]. In addition to the increased peak data rates, carrier aggregation also allows advanced features such as multi-carrier scheduling, carrier load balancing, quality-of-service (QoS) differentiation, interference coordination, and heterogeneous deployment to be used to further increase the spectral efficiency of the system. For instance, with QoS differentiation, different subscription classes can be created whereby users are assigned a bandwidth and a preferred carrier on the basis of their level-of-service agreement. Multi-carrier scheduling can also be used to schedule users in a carrier that is experiencing less interference, thus improving throughput. Similarly, carrier aggregation can be used with inter-cell interference coordination techniques to ensure that users are scheduled in a manner that will generate less interference with surrounding cells. This is beneficial in a heterogeneous deployment where cells of different power levels and coverage areas are supported. For example, different carriers can be assigned to different coverage areas via soft reuse of the carriers. In addition, different eNB types (e.g. macro-cell, pico-cell, and femto-cell) may be assigned different carriers to avoid interference.

Naturally, carrier aggregation is designed to be backward compatible and will allow the operator to provide additional capacity without adverse impact on legacy LTE users. An operator can combine existing LTE spectrum that may be fragmented to provide larger bandwidth, thus extending the lives of their legacy networks. In addition, any new spectrum subsequently acquired or obtained through re-mining (i.e. reuse of spectrum when cellular systems are decommissioned) can also be added. Existing LTE users can access the system using one of the legacy LTE carriers. LTE-A users, on the other hand, will be able to access multiple carriers simultaneously and thereby take advantage of the wider bandwidth. In LTE-A Rel-11, a new carrier type, called an extension carrier, may be supported in addition to LTE Rel-8 carriers. An extension carrier, as the name implies, operates as an extension of another carrier, and users can access this type of carrier only as part of a carrier-aggregation set. An extension carrier may, but need not, have its own control signaling. In the case that it doesn't, control signaling is provided through the primary or anchor carrier. The main use of the extension carrier is to provide service in an environment with strong interference.

Three aggregation scenarios are possible, depending on the spectrum availability of the operators – contiguous spectrum aggregation in a single

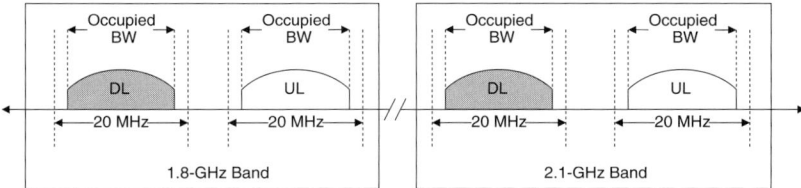

Figure 6.1. Example of carrier aggregation.

radio band, non-contiguous spectrum aggregation in a single radio band, and non-contiguous spectrum aggregation in multiple radio bands. A radio band is a part of the radio spectrum that has been reserved for a specific purpose (i.e. cellular service, public safety, etc.). For example, LTE operating band 1 is reserved for FDD mode with uplink carriers in the frequency range 1920–1980 MHz and downlink carriers in the frequency range 2110–2170 MHz [6]. Operators may have been allocated multiple carriers within the same band or across multiple bands, which can then be aggregated together. In addition, within the same band, the carriers to be aggregated may be contiguous or non-contiguous. Contiguous spectrum aggregation refers to carriers that are adjacent to each other in the frequency spectrum. An example of an FDD system with an aggregated bandwidth of 40 MHz constructed using two 20 MHz carriers is shown in Figure 6.1.

In [7], prioritized deployment scenarios for LTE-A were proposed for the aggregation possibilities outlined in the previous paragraph. These deployment scenarios were prioritized by several cellular operators to enable a co-existence and feasibility study of the carrier-aggregation feature. Some representative examples of the proposed scenarios are listed in Table 6.2 (adapted from [7]). As shown in Table 6.2, for FDD asymmetric carrier aggregation (e.g. larger bandwidth in the downlink than in the uplink) is possible. In TDD, however, the aggregation must be symmetric because the same carriers are used both for the downlink and for the uplink in a time-division-multiplexed manner. However, TDD allows asymmetric bandwidth usage through the use of different downlink–uplink time splits. Table 6.2 shows examples where existing LTE bands are aggregated (e.g. band 3 (1.8 GHz) + band 1 (2.1 GHz)) on the

Table 6.2. *Examples of LTE-A deployment scenarios for carrier aggregation*

Deployment scenario	Duplexing mode	Aggregated bandwidth	Carrier aggregation
Contiguous single band	FDD	UL: 40 MHz DL: 80 MHz	UL: 2 × 20 MHz (3.5 GHz) DL: 4 × 20 MHz (3.5 GHz)
Non-contiguous multiple bands	FDD	UL: 30 MHz DL: 30 MHz	UL: 15 MHz (1.8 GHz) + 15 MHz (2.1 GHz) DL: 15 MHz (1.8 GHz) + 15 MHz (2.1 GHz)
Contiguous single band	TDD	100 MHz	5 × 20 MHz (2.3 GHz)

UL, uplink; DL, downlink.

basis of the spectrum available to the operators. In addition, new radio bands (e.g. 3.5 GHz) are also planned, which will allow very large and contiguous aggregation. From an implementation perspective, non-contiguous carrier aggregation in different radio bands can be quite challenging since hardware configurations (e.g. antenna size, power amplifier, filters) might not be compatible among the different radio bands. In general, aggregated carriers should be in similar frequency ranges as shown in Table 6.2 in order to minimize hardware variations. As a result, contiguous carrier aggregation is the least challenging in terms of hardware implementation.

Note that Table 6.2 defines the carrier-aggregation scenario from a system perspective, i.e. the scenarios are system-wide. However, an individual UE may be assigned a different aggregation configuration that is a subset of the system-wide configuration. For instance, consider a system-wide aggregation of two 20-MHz carriers in the uplink and four 20-MHz carriers in the downlink. Some users may be assigned one 20-MHz carrier in the uplink and two 20-MHz carriers in the downlink, while other users may be assigned two 20-MHz carriers in the uplink and two 20-MHz carriers in the downlink. User-specific configurations allow advanced features such as load-balancing, interference coordination, and QoS management to be efficiently supported in the network. In addition, UEs of different capability classes can also be defined (e.g. some UEs can support only a maximum of 20 MHz in the uplink). In 3GPP, allowable system-wide and user-specific carrier-aggregation configurations will be defined by the 3GPP Radio Access Network Working Group 4 (RAN4) which is in charge of radio performance. In Rel-10, the number of configured uplink component carriers is always smaller than or equal to the number of downlink component carriers. As a result, uplink-heavy carrier aggregation will not be supported.

To efficiently support carrier aggregation, the concept of a primary cell and associated primary component carriers has been introduced. When a UE first establishes an RRC connection with the network, it is attached to only one cell which is referred to as the primary cell (PCell). The uplink and downlink carriers associated with this PCell are called the downlink primary component carrier (DL PCC) and the uplink primary component carrier (UL PCC), respectively. Additional secondary component carriers can be included, depending on the desired carrier-aggregation configuration and UE capability. The uplink primary component carrier will carry data and all control channels and information needed to support data transmission using carrier aggregation. Uplink secondary carriers will be used only for data transmission. In addition, semi-persistent scheduling can be done only on the primary downlink and uplink component carriers.

In Rel-8, the peak data rate is 299.6 Mbps in the downlink with four-layer spatial multiplexing and 75.4 Mbps in the uplink, achieved using a system bandwidth of 20 MHz. With carrier aggregation, these peak rates will increase with the aggregated bandwidth. Thus, using four 20-MHz downlink carriers, a peak data rate in excess of 1 Gbps can be achieved using carrier aggregation alone. In the uplink, however, a peak data rate in excess of 500 Mbps cannot be achieved even at an aggregated bandwidth of 100 MHz using just carrier aggregation. In this case, LTE-A uplink spatial multiplexing can be used in conjunction with carrier aggregation to increase the peak data rate beyond 500 Mbps.

As part of the feasibility study for carrier aggregation, co-existence analysis must be performed to ensure that carrier aggregation will not adversely impact the performance of other wireless systems located nearby in the spectrum. The study is done using the LTE-A system as an aggressor and another cellular system as the victim. The performance of the victim system is then evaluated to ensure that any degradation is within an acceptable limit. This analysis is to be undertaken by RAN 4, with the final results captured in a study report. For LTE, the methodology for a co-existence study is defined in [8]. For LTE-A, an initial co-existence study using an LTE-A system with carrier aggregation as the aggressor (i.e. the system generating the interference) and a legacy LTE system as the victim (i.e. the system experiencing interference) is under way. Both downlink and uplink co-existence studies are being performed. In the downlink, the interference of an LTE-A eNB with other base stations interference is being studied, while in the uplink the interference of an LTE-A UE with other mobile stations is being studied. Summary results for the co-existence performance analysis can be found in [9]. An example of a downlink co-existence scenario is illustrated in Figure 6.2, where an LTE-A system involving two 20-MHz carriers is located next to a legacy 10-MHz LTE system. The performance degradation of the victim is analyzed in terms of the adjacent-channel interference ratio (ACIR), which is the fraction of the total transmit power of the aggressor system that is experienced by the victim [9]. Analysis is done for different ACIR to gauge the performance degradation in terms of the level of interference.

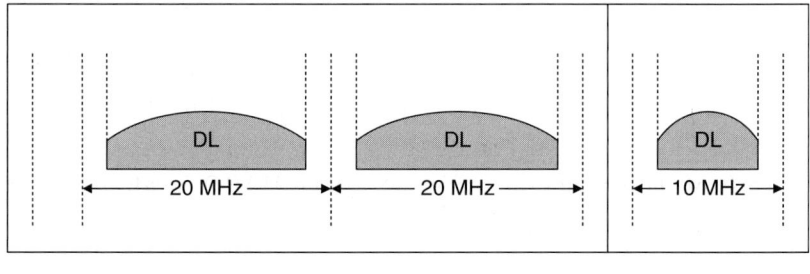

Figure 6.2. Example of LTE-A to LTE co-existence scenario.

Preliminary results have been presented in [9] both for the downlink and for the uplink. The system studied consists of five 20-MHz LTE-A carriers as the aggressor and a 10-MHz LTE carrier as the victim. For the downlink, cell throughput loss was between 5% and 14%, depending on the ACIR (the higher the ACIR, the lower the throughput loss), while cell-edge users suffer performance loss between 5% and 9%. For the uplink, both cell throughput and cell-edge throughput losses were less than 2% for all ACIR considered. Note that performance degradation is much less severe in the uplink due to the much smaller UE transmission power. From the results shown, co-existence of LTE-A with adjacent LTE is feasible for this case. Note, however, that the results are preliminary and did not consider all supported aggregation scenarios. It may be that different aggregation scenarios result in worse degradation, requiring some restriction on the aggregation scenarios that can be supported.

6.2.1 Data transmission

In LTE, data is encapsulated in a medium-access control (MAC) packet data unit (PDU) and forwarded to the physical (PHY) layer for transmission over the air. The size of the supportable transmission packet is given by the transport-block-size (TBS) table using the procedure described in Section 3.4.1.3. With carrier aggregation, however, the supportable data packet size will increase significantly. Instead of expanding the TBS table, LTE-A adopted the approach shown in Figure 6.3, whereby the physical layer remains the same as in LTE Rel-8. The MAC PDU is instead segmented into multiple packet data units such that each data packet will fit into one carrier. This interface requires minimal changes to the physical-layer specifications, and also allows individual control for the transmission of data on each of the carriers. However, separate HARQ processing and associated control signaling is required for each of the component carriers. Separate HARQ is advantageous because, if one of the segmented packets is received in error, only that packet need be retransmitted, not the entire MAC PDU. In addition, separate physical-layer processing allows individual link adaptation and MIMO support for each carrier. This can improve throughput since the

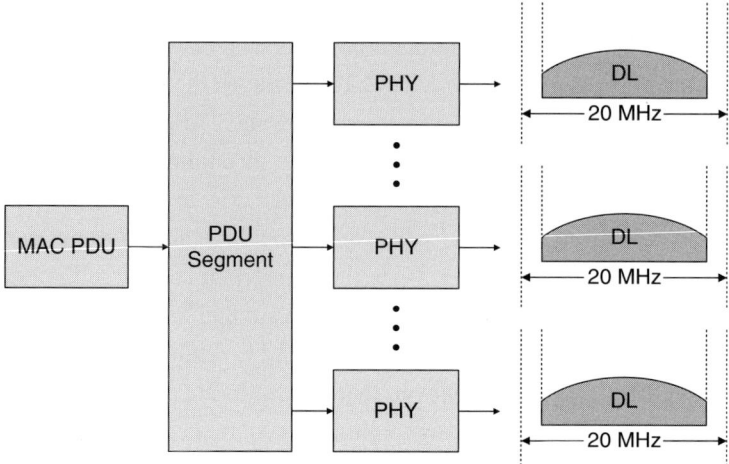

Figure 6.3. Data transmission for carrier aggregation.

amount of data transmitted on each carrier can be independently matched to the channel conditions on each carrier. However, an increase in overhead is expected due to the segmentation process. Thus, this method is not efficient for small packet size, and therefore smaller packets should not be transmitted on multiple carriers. For large packet size, however, this increased overhead is expected to be only a small fraction of the packet size.

The data transmission chain in Figure 6.3 is valid both for the downlink and for the uplink. In the downlink, OFDM is used and multiple transmitter chains will be required for non-contiguous aggregation. With contiguous aggregation, a single transmitter chain with one IFFT may be used. In the uplink, carrier aggregation is supported using N-SC-FDMA transmission. With carrier aggregation, the single-carrier property in the uplink is no longer preserved when transmitting on multiple carriers. As a result, the cubic metric increases, which requires larger back-off in the power amplifier, thereby reducing the maximum transmit power at the UE. For instance, when transmitting on two simultaneous carriers, the peak output power of the user is reduced by approximately 1–2 dB. As a result, there may be a loss of coverage of LTE-A users transmitting on multiple carriers simultaneously. This can be compensated for by smart scheduling, whereby users with poor channel conditions will be restricted to transmitting in just a single carrier.

6.2.2 Control signaling

To enable carrier aggregation, control-signaling design must be extended to provide individual control information for each carrier. In the downlink, control signaling serves several purposes – to inform the user about the size of the control region, to provide acknowledgments for uplink packet transmission, and to provide scheduling assignment in both downlink and uplink subframes. In the uplink, control signaling is used to provide feedback to the eNB, including acknowledgment for downlink packet transmission, channel-quality reports, and the rank indicator (RI), precoding-matrix indicator (PMI), and scheduling-request indicator. This control signaling must now be extended in an efficient manner to work across multiple carriers. In Rel-10, PUCCH will be transmitted only from the primary uplink component carrier.

Several challenges, however, are present in the redesign of control signaling to support multi-carrier transmission. First, carrier-aggregation configurations are user-specific and are subsets of the system-wide configuration. As a result, many different configurations must be supported simultaneously in the system, including the typical case of asymmetric aggregation where some carriers are missing their counterpart. In this case, reuse of the Rel-8 design might not be possible. Second, UE transmission is limited in the uplink and thus care must be taken to ensure that coverage is not reduced as a result of carrier aggregation. Third, some advanced features such as interference coordination may require that control be sent on a different carrier than data. While this complicates the design and introduces additional overhead, it is necessary to do so to support the use of extension carriers. Finally, overhead must be minimized in order to derive the maximum benefits. This is especially crucial in the uplink, since most of the feedback is sent on the uplink (e.g. feedback such as channel-quality reports must be sent for every configured downlink carrier).

For carrier aggregation, more changes will be required in the design of uplink control signaling than for the downlink. This is because of the transmission power limitation at the UE and the need to support user-specific aggregation configurations. In general, these configurations will be downlink-heavy (i.e. they will have more downlink carriers than uplink carriers) because typically significantly more data is consumed than generated by the UE. To support these data transmissions, feedback such as channel-quality reports and acknowledgments must be generated and transmitted by the UE. Consider the most asymmetric case of five downlink carriers and one uplink carrier. In this case, feedback information for five downlink carriers must be transmitted on just one uplink carrier. This requires the design to be highly flexible in order to address the many possible configurations and highly efficient in order to minimize power usage at the UE.

Several approaches have been offered for transmission of uplink acknowledgments (i.e. ACK/NACK transmission in response to downlink data). In LTE, this acknowledgment conveys one of three possible states per codeword – ACK, NACK, or DTX. With SU-MIMO, an acknowledgment is required for each codeword, thus increasing the number of states to five for two codewords. In carrier aggregation, the number of possibilities increases exponentially with the number of downlink carriers. For FDD, with five downlink carriers, 3125 different states are possible. This will require the equivalent of 12 information bits to transmit. Without explicit DTX support, the maximum number of acknowledgment bits can be reduced to 10. For TDD, typically 4DL:1UL is the most asymmetric configuration that will be supported in practice since 9DL:1UL will be generally used for broadcast transmission. In addition, at most five downlink carriers can be used in carrier aggregation. Thus, ACK/NACK multiplexing for TDD should be designed to carry acknowledgments from at most five downlink carriers in 4DL:1UL configuration. In this case, 40 bits will be required, given two codewords and five downlink carriers. To efficiently transmit this information, several approaches were studied. They include bundling, multi-code transmission, code selection, multi-code transmission, spreading-factor reduction, and higher-order modulation. Bundling reuses the TDD concept whereby all the acknowledgments are bundled or combined together. This requires the least power, but may be very inefficient since only one bundled acknowledgment is received for all downlink transmissions. In TDD, transmission errors are correlated (i.e. if one packet is in error, the other is also usually in error) since the radio conditions are similar among all transmissions, so bundling is a good approach. However, in multi-carrier transmission, radio conditions can be quite different across carriers. Thus, errors are unlikely to be correlated and bundling across carriers can degrade performance significantly. Multi-code transmission allows the UE to transmit using multiple codes simultaneously. This allows acknowledgments to be multiplexed together in the code domain at the expense of higher transmission power. Code selection allows the UE to select one of multiple reserved codes to convey the information. This method uses less power, but requires a large overhead since a large number of codes (analogous to a large number of uplink resource blocks) will have to be reserved. Thus, code selection works well when the number of ACK/NACK transmissions is small, but requires substantial resources for a large number of bits. It may also be possible to reuse the Rel-8 PUCCH format 2 (used for CQI/PMI/RI transmission) for acknowledgments. Unfortunately, no single method is universally superior. As a result, different methods have been defined according to the number of acknowledgments to be transmitted.

For FDD, a maximum of 10 bits can be used to transmit acknowledgments for carrier aggregation, depending on the configuration. Two different PUCCH formats will be used. For up to four acknowledgment bits, PUCCH format 1b with channel selection will be used. This is similar to the scheme used for TDD in Rel-8 (see Section 4.4.1). For 5 to 10 acknowledgment bits, however, a new PUCCH format, namely format 3, which is based on DFT-S-OFDM, will be used. A detailed description is shown in Figure 6.4. With this format, the acknowledgment bits are first encoded into 48 encoded bits. The encoded bits are next scrambled and then mapped to 24 QPSK modulation symbols. The modulation symbols

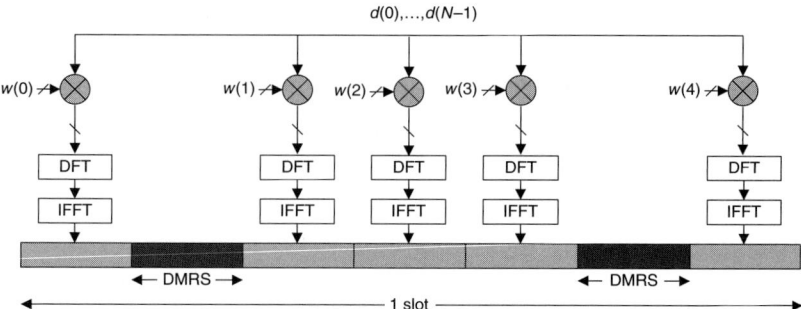

Figure 6.4. PUCCH format 3.

then undergo orthogonal spreading and then transform coding before mapping to the resource elements.

From Figure 6.4, it can be seen that this format is similar to PUCCH format 2, with the main differences being the coding rate and the number of users supported. For example, with 10 acknowledgment bits, PUCCH format 2 provides a coding rate of 0.5, while PUCCH format 3 provides a coding rate of 0.21. This reduced coding rate results in a performance gain of approximately 2 dB, which was the main reason why this new format was introduced. However, a maximum of five users can be supported in this format, compared with a maximum of eight users using PUCCH format 2.

When PUCCH format 1b with channel selection is used, resource selection can be done implicitly as in Rel-8. However, with PUCCH format 3, implicit resource selection is no longer viable due to the large overhead associated with format 3. In this case, a UE will select a PUCCH resource to transmit its acknowledgment on the basis of an RRC-configured resource and a new field called the ACK/NACK resource indicator (ARI) given in the downlink scheduling assignment. The combination of these two fields allows low overhead while providing some scheduling flexibility in resource assignment. In this case, a UE is assigned a specific PUCCH resource via RRC signaling. However, some limited adjustment to the final resource selection can be made dynamically via the ARI. This allows the same resource to be assigned to multiple users, thus minimizing overhead. In cases of conflict where users with the same PUCCH resource are scheduled in the same subframe, the ARI can be used to avoid this contention.

For TDD, a maximum of 20 bits can be used to transmit acknowledgments for carrier aggregation, even though 40 bits are needed to support the most severe configuration. Spatial bundling across codewords will be done. That is, within each subframe, acknowledgments for the two codewords are bundled together to form one decision. For up to four acknowledgment bits, PUCCH format 1b with channel selection will be used. For from 5 to 20 acknowledgment bits, however, PUCCH format 3 will be used.

In addition to the uplink acknowledgments, channel state information such as channel-quality reports, rank reports, and PMI reports must also be transmitted. In LTE-A, this information can be sent periodically or upon request by the eNB. In Rel-10, the PUCCH will be transmitted only from a single uplink component carrier. As a result, the CQI/PMI/RI reports from up to five downlink carriers will need to be transmitted from just one uplink carrier. Currently, an independent CQI/PMI/RI configuration for each component carrier (i.e. cycling through each carrier) is supported as a baseline. To provide maximum configurability, additional restrictions such as a minimum reporting period should not be imposed, and therefore it is up to the eNB to manage the reporting configuration to avoid collisions. For carrier aggregation, a set of such reports will be required for each of the configured downlink carriers. This may be done, for instance, periodically in a time-division-multiplexed manner whereby reports for different carriers are sent at different times. Different carriers may be configured with different reporting periods and types to save overhead. For instance, subband CQI reporting may be configured on the primary downlink carrier with only wideband reporting configured for secondary carriers. The eNB may also rely mostly on aperiodic CQI reports that require only infrequent periodic wideband reports. With an independent CQI/PMI/RI configuration for each carrier, it is possible that collisions between reports from different carriers may occur. Although the eNB can try to avoid collisions, this is not always practical, especially in TDD systems, where the number of uplink subframes may be restricted. In addition, with concurrent SRS transmission, finding a configuration that completely avoids collisions can be quite complicated. In the case of collision, only the report with the highest priority is sent. The other reports are dropped. This allows greater flexibility in the configuration of each carrier, especially in selecting the periodicity of the CQI/PMI/RI reports.

In the downlink, the main design challenge for control signaling is the transmission of scheduling assignments to the UE. For carrier aggregation, separate scheduling assignment grants, such that each grant provides information about data transmission in one carrier, have been adopted. Thus, if a user is scheduled to receive data on two carriers, two downlink scheduling grants will be given. Similarly, if a user is scheduled to transmit on two carriers, two uplink scheduling grants will be provided. Using separate grants means that the eNB has the ability to perform dynamic grant load balancing and interference coordination among the component carriers on a subframe basis. To allow scheduling flexibility, cross-carrier scheduling such that a grant for data transmission in one carrier is given in another carrier is allowed. Thus, a new field called the carrier indication field (CIF) will be added to the scheduling grant to inform the UE of the carrier index the grant applies to.

In the downlink, the size of the control-channel region (given in terms of the number of OFDM symbols) can be dynamically changed on a subframe basis. This information is conveyed in the PCFICH, and users must detect this information on all of the carriers. However, an error situation may arise with cross-carrier scheduling when the user correctly received a grant in one carrier but cannot successfully decode the PCFICH of the target carrier. Several proposals were put forward regarding the size of the control region across the multiple carriers. One proposal was to fix the size of the control region to be the same across all carriers. This facilitates cross-carrier scheduling since users are aware of the size of the control region in all carriers. A counter-proposal was to independently set the size of the control region per carrier. This allows more optimal tuning of resources per carrier at the expense of having the user independently decode the PCFICH in all carriers. This approach was adopted since it allows the most flexibility and can result in the lowest overhead. However, to eliminate potential error when cross-carrier scheduling is used, the user will also be informed of the size of the control region in the other carriers that it is scheduled.

The downlink acknowledgment channel structure for carrier aggregation is based on Rel-8 design. The PHICH is transmitted on the same downlink carrier as the uplink scheduling assignment. In general, resource-index selection should reuse LTE Rel-8 design as much as possible to avoid introducing a new method [10]. For the PHICH, Rel-8 design can be extended to efficiently support system and UE-specific carrier-aggregation configurations. Two main issues were considered – which downlink carrier to send the PHICH on, and how to select the PHICH resource in that carrier. For PHICH carrier association, two approaches have been considered so far. The first is to explicitly define the downlink–uplink pair. For asymmetric carrier aggregation, several uplink carriers can be paired with one downlink carrier. The second approach is to transmit acknowledgment on the same downlink carrier as the uplink scheduling assignment. This solves the issue with uplink-heavy aggregation and also does not require an explicit pairing relationship to be defined. Owing to the potential need to support many simultaneous UE-specific configurations, the second approach was adopted due to the decoupling of aggregation configuration with the PHICH. As for the problem of PHICH resource selection, in Rel-8 the number of PHICH indices is configurable, with the maximum number equal to roughly twice the number of downlink resource blocks. In most cases this is sufficient even when MU-MIMO and uplink-heavy aggregation is considered. For instance, in a 20-MHz system, 200 unique PHICH resources are available. This means that potentially up to 200 transmissions can be acknowledged in one downlink carrier. In general, the system is unlikely to be fully loaded and the number of scheduled users is limited by the size of the control channel. Thus, the number of unique PHICH resources required per subframe is likely to be significantly less than the PHICH capacity. Thus, it is seen that current Rel-8 PHICH provisioning should provide enough capacity to support carrier aggregation.

Since in Rel-10 uplink-heavy carrier aggregation will not be supported, PHICH resource collision can occur only as a result of cross-carrier scheduling. For dynamically scheduled transmission, the DMRS cyclic-shift mechanism can be used by the eNB to avoid collisions. Nonetheless, for semi-persistent scheduling, collision remains a possibility. However, it was agreed that, for Rel-10, SPS can be done only on the primary component carriers. As a result, cross-carrier SPS will not be supported. Thus, PHICH collision is no longer an issue and no additional standardized mechanism is needed for Rel-10 for PHICH resource selection.

6.3 Downlink multi-antenna transmission

In LTE Rel-8, downlink data transmission using a maximum of four data layers (a data layer is equivalent to a data stream) to the same user (4×4 SU-MIMO) is supported. A summary of downlink transmission modes supported in Rel-8 is outlined in Table 5.1. Four different multi-antenna transmission techniques are supported – transmit diversity (transmission mode 2), closed-loop spatial multiplexing using precoding vectors (transmission modes 4, 5, and 6), open-loop spatial multiplexing (transmission mode 3), and user-specific beamforming (transmission mode 7). Closed and open-loop spatial multiplexing requires the use of common reference signals (CRSs). The number of required CRS sets grows linearly with the number of antenna ports. Thus, for four-layer transmission (transmission mode 4), four sets of CRSs are needed, which can result in high overhead. User-specific beamforming, on the other hand, uses a fixed number of dedicated reference signals regardless of the number of antennas. However, in Rel-8, only one data layer can be transmitted using user-specific beamforming transmission mode 7.

In LTE Rel-9, a new downlink transmission mode (transmission mode 8) is introduced, wherein the dedicated reference-signal (DRS)-based beamforming is extended to a maximum of two data layers. Transmission modes 7 and 8 have been optimized for TDD, and are being deployed by major TDD operators around the world. In Rel-10 LTE, the downlink spatial multiplexing is extended to support eight data layers (8×8 MIMO), thus increasing the peak data rate by a factor of 2 over LTE Rel-8 for single-carrier transmission. In Rel-10, a new PDSCH transmission mode (transmission mode 9) that has been optimized for both FDD and TDD is introduced. If a CRS is used to support this, the overhead will increase substantially. To avoid this, a DRS has been adopted for this new mode. In the following section, the details of LTE Rel-9 and Rel-10 downlink spatial multiplexing (transmission modes 8 and 9) are described together with the performance results.

6.3.1 LTE Rel-9 downlink spatial multiplexing

The downlink spatial multiplexing scheme in Rel-9 is mainly geared towards TDD since the spatial channel information is obtained at the base station through a sounding reference signal (SRS) using the principle of channel reciprocity. The user-specific reference-signal-based single-layer beamforming scheme in Rel-8 is extended in Rel-9 to provide either one or two streams (layers) of data to a single UE using DRS (SU-MIMO) or two layers of data to two UEs (one layer each) using the same time–frequency resource (MU-MIMO). In summary, the Rel-9 downlink spatial multiplexing scheme supports a single transmission mode for SU-MIMO (ranks 1 and 2) and MU-MIMO with dynamic transition between SU-MIMO rank 1, SU-MIMO rank 2, and MU-MIMO. It is also possible to extend MU-MIMO to four streams with a non-orthogonal DRS. The new MU-MIMO implementation in Rel-9 provides a significant enhancement compared with Rel-8 MU-MIMO (transmission mode 5). This is because Rel-8 MU-MIMO does not provide any performance gains compared with SU-MIMO or single-layer beamforming since it is limited by coarse quantization (due to the use of a codebook) and the lack of interference suppression at the UE.

In order to support dual-layer beamforming transmission, two layers of user-specific reference symbols need to be used to demodulate the two streams of data to a single user or a single stream of data to two separate users using the same time–frequency allocation. The user-specific reference-signal structure is shown in Figure 6.5, where it occupies 12 resource elements per resource block, amounting to 8.3% overhead. The two layers of user-specific reference signals are overlaid on top of

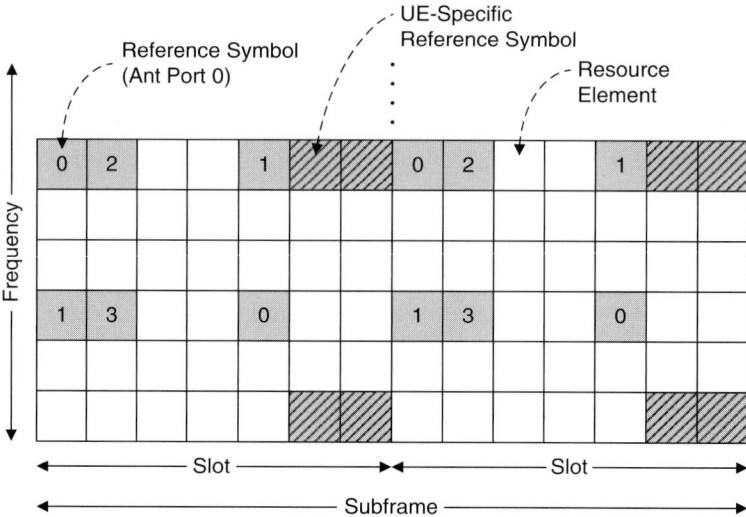

Figure 6.5. UE-specific reference signal for transmission-mode 8 dual-layer beamforming.

each other and are separated by a length-2 orthogonal cover code. The UE, after subtracting out its channel estimate, may estimate a covariance matrix representing the combined interference from a co-scheduled UE and other cell transmissions. This feature can be used by a receiver to significantly suppress the interference due to MU-MIMO. It may also be noted that the UE is not explicitly informed of the presence of a co-scheduled UE, either for purposes of feedback or for demodulation when in MU-MIMO mode. In this mode, the UE feeds back the CQI on the basis of transmit diversity. The modulation, coding, and rank for transmission to the UE are determined at the base station. As indicated before, this mode is suitable for LTE TDD where SRS is used to estimate the covariance matrix at the base station. The transmit weights are then derived from the covariance matrix. For an FDD system, the translation of uplink covariance matrix to downlink covariance matrix is possible under some channel conditions. However, this results in approximately 15%–20% degradation in system performance.

Since both SU-MIMO and MU-MIMO can be supported in transmission mode 8, the scheduler algorithm needs to be modified to include a selection criterion between the two techniques. A typical scheduler criterion based on a proportionally fair scheduling metric is as follows.

Let S_i be the estimated throughput for user i using SU-MIMO, let (M_k, M_j) be the estimated throughput for user pair (k, j) using MU-MIMO, let R_i be the average rate for user i using SU-MIMO up to that frame, and let (R_k, R_j) be the average rate for user k and j using MU-MIMO up to that frame. If the metric

$$\left(\frac{S_i}{R_i}\right) \geq \left(\frac{M_k}{R_k} + \frac{M_j}{R_j}\right)$$

then the scheduler decides to implement SU-MIMO on that frame with user i (and the associated number of streams) that maximizes the SU-MIMO throughput metric, otherwise MU-MIMO is used with the user pair (k, j) which gives the highest MU-MIMO throughput metric.

Next, the system performance of transmission mode 8 is compared with that of transmission mode 7 for two, four, and eight antennas with various numbers of transmit antennas and polarizations (uniform linear array (ULA) and cross-poles) using the urban micro-cell spatial channel model at 3 km/h, wideband CQI, and transmit weights derived from wideband SRS as summarized in Table 6.3.

The following conclusions can be drawn from Table 6.3. In transmission mode 7, an improvement in sector throughput of the order of 10% for cross-polarized antennas can be achieved when the number of cross-polarized antennas is increased from four to eight. Similarly, the fifth-percentile cell-edge throughput exhibits an improvement of approximately 19%. With four transmit antennas, transmission mode 8 performs approximately 25% better in sector throughput than transmission mode 7 using ULA antennas, while it is 15% better using cross-pole antennas. However, with eight cross-pole transmit antennas, transmission mode 8 provides 33% greater sector throughput than transmission mode 7, while the edge throughput is 8% better. In summary, the performance of Rel-9 transmission mode 8 is superior to that of Rel-8 transmission mode 7, and the improvement is more pronounced as the number of transmit antennas is increased.

Table 6.3. *PDSCH performance for various antenna configurations at the eNB (10-MHz TDD, configuration 1)*

Number of antennas and configuration	Transmission mode 7		Transmission mode 8	
	Sector spectral efficiency (bps/Hz per sector)	Edge spectral efficiency (bps/Hz per sector)	Sector spectral efficiency (bps/Hz per sector)	Edge spectral efficiency (bps/Hz per sector)
2, cross-poles	1.6	0.05	1.7	0.05
4, ULA	2.0	0.08	2.5	0.08
4, cross-poles	1.9	0.07	2.1	0.07
8, ULA	2.2	0.10	3.2	0.11
8, cross-poles	2.1	0.09	2.8	0.09

6.3.2 LTE Rel-10 downlink spatial multiplexing

In LTE Rel-10, the downlink spatial multiplexing scheme is extended to support up to eight streams, thus enabling 8×8 MIMO. Similarly to LTE Rel-9, the scheme is based on dedicated reference signals but is optimized for both FDD and TDD. Hence, eight layers of user-specific reference signals are introduced. As discussed earlier, the overhead due to supporting eight layers of user-specific reference signals is much lower than that for using eight layers of CRSs since the user-specific reference signals are more easily adaptable to the number of layers of data transmission and the size of the resource allocation. Furthermore, to support CQI and to compute channel spatial information for up to eight layers, an additional reference signal called the channel state information reference signal (CSI-RS) was introduced. The CSI-RS is transmitted sparsely and the overhead is around 0.5% considering all eight antenna ports. The maximum number of codewords or transport blocks transmitted over eight layers will remain two, with support for separate modulation and coding schemes and separate HARQ acknowledgments on each codeword.

The CSI-RS is designed such that it satisfies the following properties: it is scalable up to eight antennas, the total overhead is less than 0.5% and it is transmitted sparsely (e.g. once per radio frame), and there is good inter-cell interference rejection which means that at least three cells have orthogonal or quasi-orthogonal CSI-RSs. The inter-cell orthogonality is beneficial since it allows better CSI estimates at the UE. In addition, the CSI-RS should not be transmitted on OFDM symbols with control information or on OFDM symbols with LTE Rel-8 CRS. The CSI-RS port multiplexing for each pair of CSI-RS ports is based on CDM. With 2, 4, and 8 antenna ports, the CSI-RS can be transmitted in reuse patterns of 20, 8, and 5, respectively, thus providing CSI-RS orthogonality across five cells. This will enable a UE at the cell edge to measure CSI-RSs transmitted from adjacent cells for coordinated multi-point transmission support.

In Rel-10, the UE-specific reference-signal design is extended to support up to eight streams, which is an extension from two-stream UE-specific reference signals for Rel-9. The UE-specific reference signal also supports four users for MU-MIMO using orthogonal code cover of size four. Ideally, a covariance-matrix estimate of the downlink channel provides the best multi-rank precoder information to the eNB. This provides the eNB with the flexibility to decide the rank, MCS, and MU/SU transmission for a UE. This kind of information also maximizes the benefit of UE-specific reference signals where the eNB has the freedom to choose transmit weightings. Computing the covariance matrix at the eNB is natural for an LTE TDD system where such a matrix is calculated from the SRS utilizing reciprocity.

In Rel-10, the rank-2 and rank-4 precoding codebooks are based on the Rel-8 rank-2 and rank-4 codebooks, respectively. For eight transmit antennas, a double codebook structure for ranks 1 to 8 is introduced. This is given by $W = W_1 W_2$, where W_1 is block-diagonal and contains wideband/long-term spatial channel information, and W_2 contains subband spatial channel information. The key idea is to provide two different feedback overheads and rates for the long-term and short-term components while at the same time reducing the overhead. The block-diagonal structure of W_1 is matched to the covariance structure of a cross-polarized transmit antenna array. The rank-8 precoding codebook details are described in [11]. For rank-2 and rank-4 transmission, the Rel-8 codebooks are retained. For rank-8 transmission, the Rel-8 PMI reporting using PUCCH and PUSCH

Table 6.4. *Rel-10 PUSCH CQI modes*

CQI/PMI mode	CQI	W_1	W_2
1-2	Wideband CQI for entire system bandwidth	Single W_1: one for the entire system bandwidth (wideband)	Subband PMI W_2
2-2	Wideband CQI for the entire system bandwidth + "M-preferred" CQI (for UE-selected bands)	Single W_1: one for the entire system bandwidth (wideband)	Wideband PMI W_2 + "M-preferred" PMI W_2 (for UE-selected subbands)
3-1	Subband CQI	Single W_1: one for the entire system bandwidth (wideband)	Wideband PMI W_2

is extended to signal W_1 and W_2. As an example for PUCCH reporting mode 1-1, W is determined from two subframe reports conditioned upon the latest rank-indication (RI) report. In the first report, the RI and W_1 are jointly coded, whereas in the second report the wideband CQI and wideband W_2 are jointly coded. Similarly, the Rel-8 PUSCH CQI modes are extended to support Rel-10 CQI modes as shown in Table 6.4.

In Rel-10, downlink data transmission can be sent on eight antenna ports using a user-specific reference signal. However, related items of control information (e.g. PDCCH, PHICH, PCFICH) are still transmitted on just four antenna ports using CRSs. This is because the overhead required to support eight-antenna-port transmission for control information is prohibitive. As a result, an antenna-mapping technique must be used to map the control signals coming out of the four antenna ports to the available eight physical transmit antennas. This technique must be transparent to the user to allow proper decoding. Examples of available techniques include cyclic-delay diversity and precoding-vector switching.

6.3.3 Coordinated multi-point transmission

The motivation of the coordinated multi-point transmission (CoMP) feature is to provide air-interface support to enable cooperation among eNBs that may, but need not, be co-located (for CoMP an eNB may be a different sector within one physical eNB or may be completely different eNBs that are widely separated). Multiple eNBs may cooperate to determine the scheduling, transmission parameters, and transmit antenna weights for a particular UE. This cooperation will depend on a high-capacity backhaul link between eNBs. Closed-loop beamforming or precoding-based trans-missions will be supported in CoMP. This framework is depicted in Figure 6.5, where three eNBs may coordinate to create a multi-point trans-mission to UE1 (served by eNB1) and UE2 (served by eNB2). The objective

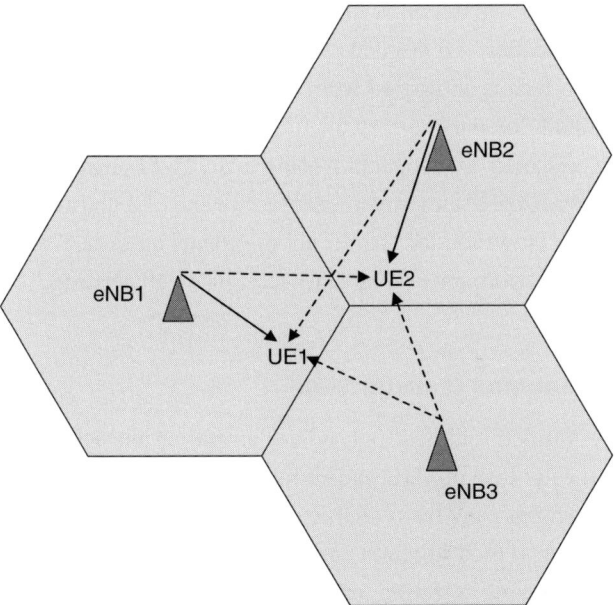

Figure 6.6. A CoMP framework for downlink transmission where eNB1, eNB2, and eNB3 can coordinate and create a multi-point transmission to UE1 and UE2.

of CoMP is to reduce interference for the UEs in the network which are close to multiple eNBs and therefore experience an interference-limited environment. The interference experienced by these UEs may be reduced and can be predicted if there is some coordination between the interfering eNBs and the serving eNB. The control overhead, feedback methods, and scheduling restrictions necessary to enable CoMP are currently being studied in 3GPP. Within the timeframe of Rel-10 a standardized interface of direct communication between two eNBs will not be re-defined and no additional features will be defined to support inter-site CoMP. Therefore, the coordination needed for CoMP will depend on proprietary interfaces that may easily be produced for co-located inter-sector coordination.

CoMP operation can be divided into two classes – coordinated scheduling/beamforming (CS/CB) and joint processing (JP). With CS/CB, data is transmitted from the serving eNB but the transmit weights, MCS selection for the UEs etc. are based on coordination from multiple eNBs. In the JP scheme, data is transmitted from multiple eNBs to a single UE so as to improve the received signal quality and cancel out the interference from other UEs. It is similar to the MBSFN concept, except that the weights, MCS, and resource allocation are derived jointly on the basis of feedback from the UE, which in turn is computed from multiple base stations.

From the system study of inter-sector CoMP (both co-located and non-co-located sites are assumed to cooperate) using CS/CB for macro-only deployments, there is a reasonable improvement in sector and edge throughput as outlined in Table 6.5. However, for HetNet environments inter-site CoMP may provide significant performance advantages (e.g. pico-cell clusters).

6.4 Uplink multi-antenna transmission

In LTE, the UE is equipped with just one power amplifier and RF transmitter chain. This means that the UE can transmit on just one antenna, although multiple antennas may be available. The eNB has the ability to direct the UE regarding which antenna to use for uplink transmission. This feature is called antenna selection, and can be used to provide some gain in transmit diversity due to the property that transmission signals from different antennas experience different radio channels. The UE can

Table 6.5. *Performance comparison of Rel-9 vs. CS/CB for a 4 × 2 system*

Scenario	ULA 0.5λ at eNB and UE		XPOL 0.5λ at eNB and UE	
	Gain in cell-average SE	Gain in cell-edge SE	Gain in cell-average SE	Gain in cell-edge SE
Rel-9 SU-MIMO (sounding)	0 (baseline)	0 (baseline)	0 (baseline)	0 (baseline)
Rel-9 MU-MIMO (sounding)	+30.18%	+23.41%	+16.74%	+13.29%
SU-MIMO + CS/ CB (sounding)	+5.73%	+11.31%	+4.94%	+11.15%
MU-MIMO + CS/ CB (sounding)	+33.73%	+34.10%	+19.10%	+21.73%

SE, spectral efficiency.

also autonomously select which of the transmit antennas to use. For LTE-A, up to four power amplifiers and RF transmitter chains will be supported. Thus, the UE can transmit simultaneously on up to four transmit antennas. Similarly to the downlink, several multi-antenna transmission modes will be supported, depending on deployment scenarios, control or data transmission, and user configurations. However, several additional challenges must be overcome for multi-antenna transmission at the UE. First, the UE is more limited in transmission power and processing capabilities, thus power consumption and algorithm complexity must be carefully managed. For example, if the UE in LTE-A is limited to the current allowable maximum transmit power of 23 dBm, power must be shared across all the amplifiers. First, this means that, per antenna, some receiver functions such as channel estimation and frequency tracking will degrade due to there being less power per antenna. Second, due to the small size of the UE, high transmit antenna correlation, which limits diversity gain, may be the norm in some cases. So the techniques must work well with various antenna correlations. Furthermore, a power

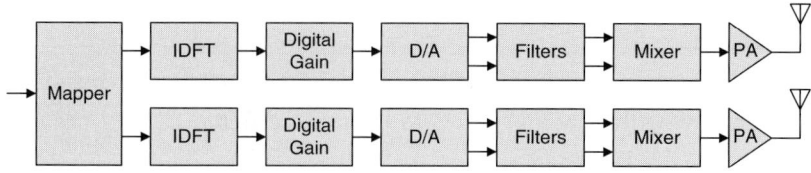

Figure 6.7. Example architecture of UE with two transmit antennas.

imbalance between different transmit antennas may be present due to UE orientation or how the handset is held [12]. This can reduce both the transmit power and the diversity gain.

With the introduction of multiple transmit antennas, changes in UE architecture will be required, including multiple radio-frequency transmitter chains and power amplifiers. This is a matter of implementation, but an illustrative example for a UE with two transmit antennas is shown in Figure 6.7. However, from a specification point of view, it is important to know the general architecture in order to intelligently design multi-antenna schemes. Examples of the proposed architectures for UE with two transmit antennas include two at 20 dBm (i.e. two power amplifiers each capable of delivering 20 dBm of transmit power), two at 23 dB, and 23 dBm + 20 dBm. Currently, two at 20 dBm is the baseline architecture for UE with two transmit antennas. This means that the user is capable of delivering 23 dBm total power when transmitting on two antennas, but only 20 dBm when just one of the antennas is used.

In general, different multi-antenna techniques are available for control and data transmissions. This is because of the different requirements and transmission modes for the two types of transmission. Generally, multi-antenna transmission for control is done to increase the reliability of the signal. For data, however, it can be used to increase either the signal reliability or the data rate. This section discusses the different transmission modes expected to be supported in the uplink for LTE-A.

6.4.1 Control channels

Transmit diversity can be used for the control channels to improve the received signal quality of these channels. Improving the performance of the control channels can result in an increased coverage area for the same transmit power or in the same coverage area for a reduced transmit power. Reducing the transmit power will lower the interference with other cells, thus improving their performance. In LTE, two different control-channel formats are used – format 1/1a/1b, which is used to transmit acknowledgments and scheduling-request indicators (see Section 4.4.1), and format 2/2a/2b, which is used to transmit channel-state information (see Section 4.4.2). Frequency diversity is already supported for control transmissions through the use of slot-hopping. Several requirements were defined for the design of transmit-diversity techniques. First, with the introduction of multiple transmit antennas, it is desirable to maintain the same cubic metric as for Rel-8 transmission if possible, or to select a method that results in a low cubic metric in order to ensure that the coverage is not severely impacted. Second, the technique must provide good diversity gain, work well with correlated antennas, and be robust against implementation non-idealities and channel variations. Third, the increase in eNB and UE complexity should be minimized. Fourth, backward compatibility is greatly desired so that LTE-A and LTE users can share the same time–frequency resources. In addition, the technique should be applicable to both two and four transmit antennas. Several techniques were investigated, including the following.

- Cyclic delay diversity (CDD). With CDD, transmissions from different antennas are delayed in time in order to introduce temporal diversity gain. This method has a low cubic metric similar to that for single-carrier transmission. In addition, it can be used without any need for orthogonal reference signals on each antenna. Because of this, CDD can be implemented in a manner transparent to the eNB, and the eNB is not required to know the number of transmit antennas being used at the UE. This allows, for example, the UE to autonomously change the number of transmit antennas without having to inform the eNB. This method can be used for all control formats as well as for data, and provides reasonable diversity gain. However, CDD performance degrades under high transmit antenna correlation. In addition, the incremental gain becomes limited as more transmit antennas are used.

- Space orthogonal-resource transmit diversity (SORTD). In this technique, users spread the information using different orthogonal resources across antennas. It can be used for all control-channel formats and provides large diversity gain. In addition, with this technique, legacy LTE users can also be multiplexed into the same time–frequency resource. Thus, it can provide backward compatibility as well. However, SORTD requires an orthogonal reference signal on each antenna, which will significantly increase the reference-signal overhead. For instance, two different cyclic shifts of the reference signals are required with two transmit antennas, thus doubling the overhead.
- Precoding-vector switching (PVS). In PVS, different precoding vectors are applied at different times to provide diversity gain. This technique is currently used on the downlink to provide transmit diversity for the synchronization signals. Similarly to CDD, this technique does not require orthogonal reference signals on each antenna and provides a low cubic metric. However, the diversity gain achieved using this technique is low, and it is not generally competitive with other techniques.
- Space frequency block code (SFBC). Several flavors of SFBC are available, including some that provide a similar cubic metric to that of LTE. SFBC is currently used to provide transmit diversity on the downlink control channels, and the basic concept is described in Section 5.2.1. This method provides a large diversity gain and has minimal complexity. However, it is not backward compatible, and thus LTE-A and LTE users cannot be multiplexed together on the same time–frequency resources.
- Space time block code (STBC). STBC is similar to SFBC with the exception that the block coding is performed across SC-FDMA symbols instead of across frequencies. This method requires an even number of SC-FDMA symbols in order to form pairs. However, for a normal cyclic prefix, seven SC-FDMA symbols are available in each slot. Hence, a different method must be used for the last SC-FDMA symbol of the slot, which reduces the diversity gain. In addition, STBC performance degrades under high Doppler shift due to the rapid changes in the propagation conditions. Finally, not all control-channel formats can use this technique.

- Frequency switch transmit diversity (FSTD). FSTD provides diversity by transmitting different groups of resource elements on different antennas. It doesn't require an orthogonal reference signal on each antenna, thus there is no increase in reference-signal overhead. A moderate performance gain can be expected with this technique. However, the increase in the cubic metric may be high, and the method is not completely backward compatible.

In LTE-A, SORTD has been selected for PUCCH format 1/1a/1b using two transmission antennas. This technique was selected due to its good performance and backward compatibility. When four antennas are available, the antennas are divided into two sets of two antennas, and SORTD is separately applied on each of the two sets. For PUCCH format 2/2a/2b, the analysis is still ongoing due to the large reference-signal overhead required for SORTD and also the desire to support a larger payload size in format 2/2a/2b for LTE-A. However, as shown in [13], different techniques provide the best performance for different payload sizes, and no technique is superior in all cases.

6.4.2 Random-access channel

Users initiate access to the system by transmitting a preamble on the PRACH. Since this is the first transmission by the UE, the eNB has no knowledge of the UE class or capabilities. As a result, any transmit-diversity scheme used by the UE must be transparent to the eNB (i.e. the eNB does not need to know which scheme and how many transmit antennas the UE used). Otherwise the eNB must perform blind detection of the number of UE transmit antennas, which can be difficult for cell-edge users under low-SINR conditions. Of the techniques discussed in Section 6.4.1, CDD and PVS are multi-antenna techniques that can be implemented in a manner that is transparent to the eNB. In PVS, the UE selects a precoding vector and applies it to the preamble to generate signals over multiple antennas. Since the UE has no knowledge of which precoding vector will give the best performance, the precoding vector is generally selected in a random manner or based on a predefined pattern. Subsequent preamble transmissions then use different precoding vectors to increase the diversity gain. Although the gain from PVS is small, random-access transmission is very robust in general and PVS may be sufficient.

Cyclic delay diversity, on the other hand, provides a better diversity gain and can also be made transparent to the eNB if the delays introduced on different transmit antennas are kept to within the cyclic prefix. CDD works by introducing multiple paths through the delays, thus increasing the frequency diversity. However, the timing-estimation accuracy is degraded because the transmit power is spread among the different paths. Thus, it is not clear whether the additional diversity gain will be sufficient to overcome this loss.

As shown in Section 4.10.1, PRACH is the most robust channel with the lowest SINR requirements among all uplink channels. While multi-antenna transmission can increase the reliability of the preamble transmission, it might not result in increased reliability or coverage of the cell. Hence, multi-antenna techniques for the PRACH may remain undefined in the specification and instead be left to implementation. This is the same approach as in multi-antenna transmission of downlink synchronization signals in LTE.

6.4.3 Data channel

For data transmission on the PUSCH, two types of multi-antenna transmission technologies are being considered. The first involves techniques that will improve the performance and reliability of PUSCH transmission by improving the received SINR. They include transmit diversity and beamforming via precoding. The second involves techniques that will increase the data throughput of PUSCH transmission using the spatial multiplexing principle. In the downlink of LTE, these different techniques are associated with a transmission mode that is configurable on a per-user basis. For example, transmission mode 2 refers to transmit diversity, transmission mode 3 is open-loop spatial multiplexing, transmission mode 4 is closed-loop spatial multiplexing, and transmission mode 7 is beamforming with a user-specific reference signal.

The first type of multi-antenna transmission technique is used to improve the received signal quality of the PUSCH. In LTE-A, two techniques were investigated – open-loop transmit diversity and closed-loop precoding. Open-loop transmit diversity does not require knowledge of the channel state information. The available techniques are similar to those described in Section 6.4.1 for the PUCCH, namely CDD, PVS, STBC, and SFBC. Closed-loop precoding requires the eNB to estimate uplink channel state information, normally through the use of an SRS. Closed-loop precoding can be based either on the short-term channel state information or on long-term channel statistics [13]. Precoding based on short-term information is applicable for low-mobility users, for which such information can be applied before it becomes obsolete due to the varying channel, whereas precoding based on long-term information is applicable for high-mobility users and users that are semi-persistently scheduled. This technique is robust with respect to antenna correlation and antenna imbalance, and is applicable to both two and four transmit antennas.

An illustration of the closed-loop precoding technique is shown in Figure 6.8. The transmitted data is precoded using a precoding matrix that is selected by the eNB, and given to the UE via the scheduling grant. This precoding is based on predefined codebooks similar to those defined for the downlink. The eNB selects the best PMI from the codebooks on the basis of channel state information and informs the UE of its selection (e.g. through control signaling or RRC configuration). In this case, SRSs must be transmitted from both antennas so that the eNB can estimate the

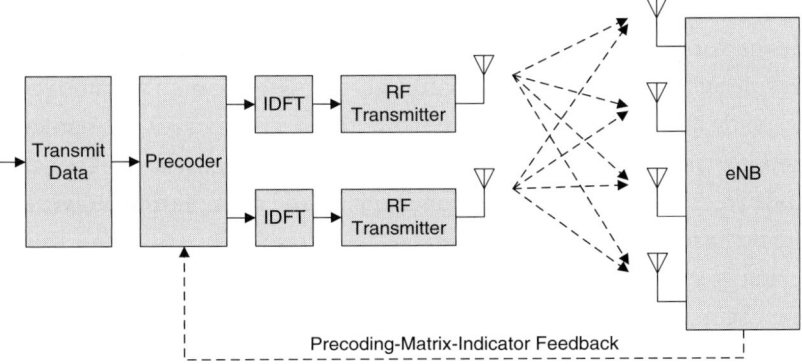

Figure 6.8. Uplink closed-loop precoding for LTE-A.

full spatial channel either on a short-term or on a long-term basis. This introduces a performance loss from uplink overhead that will have to be overcome through a precoding gain. In addition, the precoding can be done on a wideband basis (i.e. one precoding matrix applied to the entire bandwidth) or on a subband basis (i.e. different precoding matrices can be applied to different portions of the bandwidth). Subband-based precoding is more beneficial, but is more complicated and will require more control signaling. Performance analysis and complexity tradeoffs between open-loop transmit diversity and closed-loop techniques have been investigated in [13]–[14]. The general consensus is that closed-loop precoding provides better performance than does open-loop transmit diversity. This is because higher gain can be achieved with closed-loop precoding due to the ability to direct the transmitter beam from the users. However, closed-loop precoding is more complicated and requires additional overhead compared with open-loop transmit diversity.

The second type of multi-antenna transmission technique is used to increase the data throughput of the PUSCH. This is done through spatial multiplexing whereby different data streams are multiplexed into different spatial layers. In LTE, this technique is known as single-user multiple-input multiple-output (SU-MIMO). Up to four spatial layers will be supported in the uplink of LTE-A. This will increase the peak data rate for a 20-MHz carrier from 75.4 Mbps in LTE to 149.8 Mbps in LTE-A. Note that this peak rate is per carrier. When carrier aggregation is considered, the maximum peak rate will increase to 748.9 Mbps when five uplink 20-MHz carriers are used, thus exceeding the LTE-A requirement of 500 Mbps. The basic design for uplink SU-MIMO is the same as for downlink SU-MIMO. In this design, at most two codewords are supported using the same layer mapping as for the downlink. This is illustrated in Figure 6.9 for four-layer multiplexing. In this case, two of the layers are multiplexed into the same codeword. Note that different layer-mapping and multiplexing procedures are performed for layers 1, 2, 3, and 4.

In Figure 6.9, each codeword goes through the physical-layer processing chain, including coding and rate matching, scrambling, and mapping to symbol modulations. The codewords are then mapped to the different layers and undergo DFT-precoding. In LTE-A, layer shifting was studied

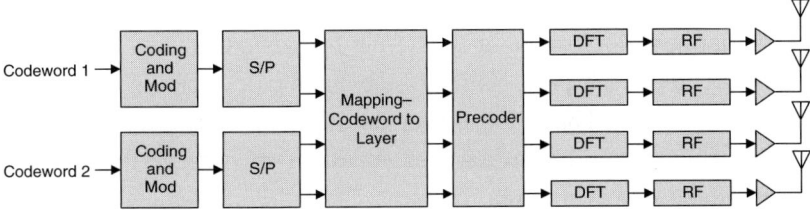

Figure 6.9. Example of four-layer spatial multiplexing.

as a method to equalize performance among the different data layers. Layer shifting is a technique in which different layers coming out from the codeword to the layer-mapping block are mixed together. This allows all the layers to be transmitted on all the antennas, thus providing similar performance across layers. This mixing can be done, for example, on an SC-FDMA symbol-by-symbol basis. In this case, different layers are mapped to different antennas in different SC-FDMA symbols. For instance, in the first SC-FDMA symbol, layers 1, 2, 3, and 4 can be mapped to transmit antennas 1, 2, 3, and 4, respectively. In the next SC-FDMA symbols, layers 1, 2, 3, and 4 can be mapped to transmit antennas 2, 4, 3, and 1, respectively. In the downlink, layer shifting is not done, and each layer experiences a different performance. System simulation results demonstrated that this layer differentiation results in higher overall data throughput than when layer shifting is used. As a result, layer shifting was not included in uplink MIMO. Next, precoding is applied, and the symbols are then mapped into different RF transmitter chains and transmit antennas.

The performance of uplink SU-MIMO depends on the design of the codebooks. Codebook design for the uplink is different than that for the downlink due to the use of the single-carrier transmission scheme. This means that the codebook should have a cubic-metric-preserving property (i.e. the cubic metric should not increase with SU-MIMO transmission). A lower cubic metric allows higher transmission power, which can enhance coverage or provide higher throughput. However, the disadvantage of this requirement is that the best-performing codebooks might not be cubic-metric-preserving ones. In addition to the cubic-metric-preserving property, high transmit-antenna correlation must be considered, so the entries in the codebook must have good separation from each other. In addition, equal transmission power per layer is required so that power balancing can be achieved. Naturally, the number of entries in each codebook should be small in order to keep the signaling overhead low.

Different codebooks are required for different numbers of transmit antennas and ranks (or spatial layers) because the size and entries of the codebooks need to be optimized individually. For example, with two transmit antennas, two codebooks – one for rank-1 transmission and one for rank-2 transmission – are needed. Similarly to the case for the downlink, codebooks will need to be defined for two and four transmit antennas. In LTE-A, codebook design for two transmit antennas has been finalized. In this case, rank-1 transmission uses the same codebook as for the downlink, whereas rank-2 transmission uses the identity matrix. The baseline codebook design for rank-1, rank-2, and rank-4 transmission for four transmit antennas has been finalized. The rank-1 codebook has 24 constant entries with 16 entries and 8 entries supporting antenna-pair turn-off. The antenna-pair turn-off entries are used to allow transmission on only one pair of the transmit antennas. This allows two of the four available antennas to be selected for transmission in the case of severe antenna imbalance. The rank-2 codebook has 16 entries consisting of cubic-metric-preserving matrices. The codebook for rank-4 transmission is the identity matrix. Note that the identity matrix is used for full-rank transmission for both two and four antennas. This means that each layer is transmitted on one antenna. This is the same as for MU-MIMO in LTE, where each of the users is considered a virtual MIMO user. The eNB can then use the same receiver as the one it uses for MU-MIMO to process full-rank SU-MIMO transmission. The rank-3 codebook design is still under study due to the non-negligible performance loss when a codebook with the cubic-metric-preserving property is used. Note that each element of the precoding matrix consists of a limited number of values in order to limit the complexity.

To support uplink spatial multiplexing, unique reference signals must be sent on each transmit antenna to enable the eNB to perform channel estimation on each transmit antenna separately. In the uplink, reference signals are user-specific (i.e. unique to each user rather than being shared).

As a result, they can be precoded (by applying the precoding matrix) in the same manner as the data. Using precoded reference signals helps save overhead since the required number of unique reference signals is based on the rank rather than on the number of antennas. For example, with four transmit antennas, only two unique reference signals are required for rank-2 transmission. If the reference signals were not precoded, then four unique reference signals, one for each antenna, would be required. Similarly to reference-signal support in MU-MIMO LTE, the unique reference signals can be generated by using different cyclic shifts. Although 12 different cyclic shifts are available per sequence, in practice only about 4 can be used in order to maintain orthogonality between the different signals.

6.4.4 Coordinated multi-point reception

Coordinated multi-point (CoMP) processing is a technique for coordination among different eNBs that can be used to improve system and user performance. Uplink CoMP is a network feature, and in most cases can be done in a manner transparent to the users. In WCDMA, coordinated reception in the form of soft handover is supported. There, information from different base stations is combined at the radio network controller. However, LTE architecture is a flat architecture that is based on the Internet Protocol, and therefore no radio network controller is present. As a result, either coordination techniques will have to be implemented in a distributed manner at participating eNBs or a centralized coordinating entity will have to be introduced. Although some techniques by which to extend CoMP to control signaling have been proposed, the main focus is on improving data throughput since control signaling is considered to be quite robust in LTE. As the name implies, CoMP requires cooperation among different eNBs that form a cooperation set. In general, the more eNBs in the cooperation set, the better the performance improvement. However, the complexity increases exponentially with the number of eNBs to coordinate. In the uplink, three different CoMP approaches are available – coordinated scheduling, coordinated beamforming, and joint processing. Coordinated scheduling requires the eNBs in the cooperation set to coordinate their scheduling decisions such that interference is limited or eliminated. For example, users in different eNBs can be scheduled on different resource blocks in order to avoid interference. Coordinated beamforming refers to coordination of the transmitted signals by the users such that either the users do not interfere with each other or the interference is minimized. Joint reception refers to joint detection of the uplink signals by multiple eNBs in order to improve detection quality. Joint reception can be used to serve a single user or multiple users simultaneously using MIMO techniques.

The cooperating eNBs themselves can be made up exclusively of eNBs at the same site (e.g. eNBs that make up the sectors of a site) or of eNBs across different sites as shown in Figure 6.10. In either scenario, the UE is still attached to just one cell, and thus is controlled by just one eNB (called the serving eNB). Same-site coordination is much easier, since the sectors are served by the same equipment, and generally does not require data transfer on an external backhaul. In addition, it can be done in a proprietary manner without standards support. However, only limited gain can be expected, especially for users that are on the cell borders with other sites. Multi-site coordination requires standardized signaling support in order for it to work across equipment from

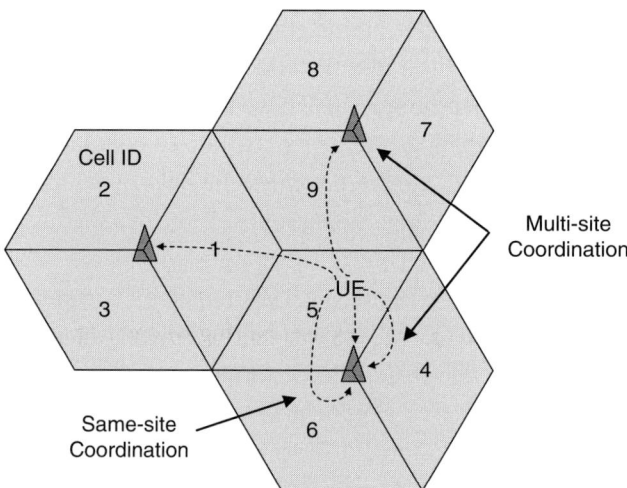

Figure 6.10. Uplink eNB coordination.

multiple vendors, large backhaul capability for data transfer, and possibly the addition of a centralized entity to perform the coordination. Significantly larger gain, however, is possible with multi-site coordination, albeit at the expense of additional complexity and backhaul costs.

Coordinated scheduling requires different eNBs to coordinate their scheduling decisions in such a way as to optimize certain performance metrics (e.g. cell throughput or average user throughput). The goal is normally to coordinate scheduling decisions such that interference is minimized. This means that scheduling decisions must be either coordinated among different eNBs in a distributed manner or obtained from a centralized entity. In general, a centralized architecture is not preferred due to the additional delay incurred and the need to introduce a coordinating entity. However, it can be done in certain scenarios, such as same-site coordination. In LTE, inter-cell interference coordination techniques can already be supported using the high-interference indicator (HII) and the overload indicator (OI). The HII provides an indication of which resource blocks in the reporting eNB will generate high interference with the neighboring eNBs, while the OI indicates which resource blocks are experiencing high interference at the reporting eNB. Through the exchange of this information, neighboring eNBs can coordinate their scheduling decisions such that interference is minimized. For instance, in lightly loaded systems, the eNBs can avoid assigning resource blocks that will contribute large interference with neighboring cells on the basis of their reported overload indicators. Another potential scheme is cooperative silencing, whereby the dominant interferer is not allowed to transmit (or allowed to transmit only at reduced power) on some set of resource blocks or in some subframes. This scheme has been shown to be beneficial in a heterogeneous deployment scenario where a low-power eNB is added into a macro-cell in order to increase coverage and performance [15]. Another example of coordinated scheduling involves user pairing where the pairing may be done across eNBs [16]–[17]. In this scheme, the eNBs have knowledge of the channel response from all users in the system and thus can pick users that are orthogonal in the spatial domain. The users are then scheduled using the same time–frequency resource with the expectation that they will create minimum interference with each others.

The main concerns with coordinated scheduling are the amount of information that must be exchanged over the backhaul and the performance and complexity of the distributed algorithm. In general, the distributed algorithm is run in an adaptive manner and may require several iterations before convergence to a good solution occurs. This increases the complexity, introduces delay, and requires scheduling decisions to be exchanged multiple times. This exchange of information is done through the backhaul, which may have high latency and limited capacity. In addition, the complexity and delay can increase substantially as the number of cooperating eNBs grows. Increasing the delay reduces the benefits since the gains from the scheduling strategies (e.g. scheduling users only when they have a good channel) may be lost due to information becoming out of date. In addition, scheduling decisions are implementation-specific and not standardized across equipment vendors. Thus, in a system that is served by multiple vendors, coordinated scheduling might not be feasible.

Joint reception (also known as macro-diversity reception) can be performed with various degrees of cooperation and using different techniques. For example, multiple eNBs may receive the same data packet from the UE, with each eNB decoding the data packet independently. The eNB that correctly decodes the packet will then forward the data to the serving eNB. Alternatively, all eNBs can forward quantized received signals to the serving eNB for final decoding. The different multiple received signals at different eNBs can then be combined at the serving eNB. This can be thought of as the serving eNB having virtual antennas made up of the physical antennas of all the eNBs. The second approach results in a greater improvement than does the first, because more sophisticated detection and interference-suppression techniques can be used, but requires significantly more information transfer between eNBs. As noted previously, coordination can be performed within the same site or across multiple sites. Since joint reception relies on receive diversity across the eNBs to improve performance, especially for cell-edge users, same-site joint reception does not provide a significant gain, due to the similar channel conditions. This is unfortunate, since same-site joint reception can be implemented very simply. Joint reception across multiple sites, however, has been shown to significantly improve performance [18]. Another example of joint reception involves the cancelling out of uplink interfering transmissions from the other cells. In this case, the eNB has some knowledge of the interfering uplink users from other cells (e.g. modulation and coding level, reference signals). It can then use this information to regenerate and cancel out the interference from the desired signal. In this scenario, each eNB must try to decode all interfering data packets. Alternatively, each eNB may first decode its own packet and, if successful, forward the decoded packet to neighboring eNBs for use in interference cancellation.

One difficulty with joint reception is the timing difference of the uplink signals arriving at all the eNBs in the cooperation set. With SC-FDMA, the uplink transmission must arrive at the eNB within a predefined timing window given by the cyclic-prefix length as discussed in Section 4.7. Otherwise, the performance of any uplink transmission that arrives outside of this window will degrade severely. The propagation time of the signal depends on the distance to the eNBs, and with multiple eNBs different propagation times will be observed. However, as long as the difference in delays is smaller than the cyclic prefix, the performance is good. Two approaches can be used to address the timing issue. The first approach is to limit the eNBs in the cooperating set to those with delays that are within the cyclic prefix. This limits the number of potential eNBs that can form the set. Different timing-advance-adjustment schemes have been presented in [19], which address this problem. The second approach is to use an extended cyclic prefix. This will allow the signals to be received by more eNBs, thus improving performance. However, using an extended cyclic prefix will reduce the number of SC-FDMA symbols per subframe by two and thus reduce throughput by approximately 14%.

Coordinated beamforming refers to coordination of the transmitted signals by the users such that the interference arising from different users is minimized. This could be implemented, for example, through the selection of proper precoding matrices (i.e. the beamforming weights) for the users to be scheduled in different cells. Alternatively, this can be done through proper user selection (i.e. selection of users that will transmit in the uplink) by the scheduler in different cells such that they create minimum interference with each other. The received signal can be decoded by the serving cell only, or jointly across multiple cells. In the case of joint reception, all the receive antennas at the different eNBs can be viewed as virtual antennas of the serving eNB. For joint reception, soft decisions must be forwarded through the backhaul to the serving eNB. In the case of reception in just the serving cell, scheduling decisions and channel state information (either short-term or long-term) must be exchanged between cells. This process can be implemented in an iterative manner. For instance, an eNB selects the user and beamforming weights and then informs the neighbors of this information. Neighboring eNBs then select their users and beamforming weights, and report this information back. The process is then iterated until an optimal set of users and weights has been found. The drawbacks of coordinated beamforming are similar to those for coordinated scheduling, namely complexity, delay, and backhaul cost.

6.5 Heterogeneous network

Traditional cellular network architecture was mainly designed for voice, and is more than two decades old. It is estimated that 64% of all mobile IP traffic will be video by 2013 [20], so it will be the single largest component of mobile data usage. The streaming video rates supported will be of the order of 600 kbps to 1.5 Mbps and will be used in mainly hotspot and indoor applications for downlink and for public-safety applications (e.g. streaming video from the scene of an incident) and/or high-speed machine-to-machine communications (e.g. video surveillance) in the uplink direction. It will be difficult to support high edge data rates and usage in an LTE macro-cellular system even with advanced multi-antenna schemes employing four transmit and receive antennas at the eNB. Recently, the concept of heterogeneous networks has been defined in the 3GPP standards. Heterogeneous networks consist of a traditional macro-cell-based network augmented with various types of low-power network nodes that address the capacity and coverage challenges resulting from the growth of data services. The traditional macro network is deployed to provide umbrella coverage and the augmenting component provides an underlay network that could consist of a new type of network node (pico, relay, remote radio head, femto and distributed antenna system) or be a complementary technology like Wi-Fi. In this section, the various low-power nodes and their performance in various deployment scenarios are discussed. Finally, various inter-cell interference coordination (ICIC) techniques for heterogeneous deployments are summarized. These ICIC techniques can be used to further enhance the performance of heterogeneous deployments.

6.5.1 Heterogeneous network overview

A heterogeneous network consists of low-power nodes underlain in a macro-cell network. The characteristics of the various types of heterogeneous nodes are summarized in Table 6.6.

An example of a traditional macro-cellular network augmented by different types of heterogeneous network nodes is shown in Figure 6.11. This figure shows a heterogeneous network consisting of below-rooftop and in-building deployments underlying a macro-cell network. The deployment of low-power pico nodes below the roof top will

Table 6.6. *Characteristics of heterogeneous nodes*

Type of node	Placement	Transmit power (W)	Number of antennas	Backhaul characteristics
Micro-cell	Outdoors	1–5	2T + 2R, 4T + 4R	Dedicated wireline
RRH node	Indoors or outdoors	1–5	2T + 2R, 4T + 4R	Dedicated wireline
Relay	Indoors or outdoors	1–5	2T + 2R, 4T + 4R	Wireless out-of-band or in-band
Pico-cell	Indoors or outdoors	0.2–1	2T + 2R, 4T + 4R	Dedicated wireline
Femto-cell	Indoors	0.1	2T + 2R, 4T + 4R	Residential broadband

T, transmit; R, receive.

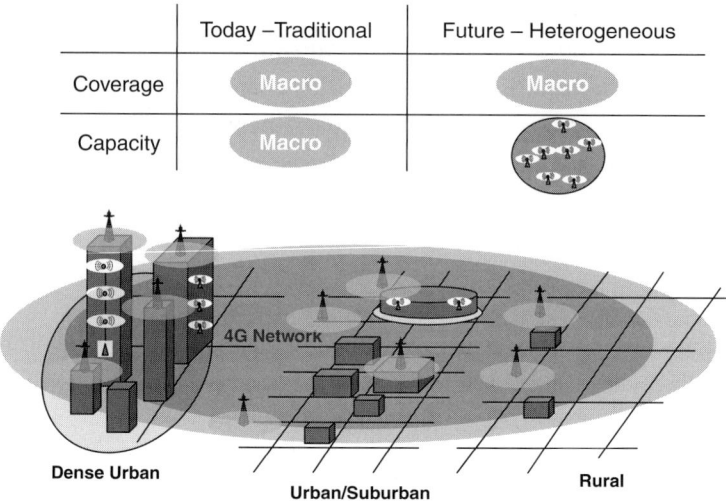

Figure 6.11. Macro-cell network with various types of heterogeneous nodes underlay.

allow more street-level coverage and increase the spectrum efficiency of the overall system by enabling an increased use of a higher-level modulation and coding scheme. The in-building coverage can be achieved using a remote radio head, a distributed antenna system, or Wi-Fi offload.

The deployment of heterogeneous nodes is challenged by the availability of backhaul and its physical size. As outlined in Table 6.6, the heterogeneous node will need to be much smaller in terms of both size and power. Further, there will be additional complexities with heterogeneous-network deployment, such as the significant increase in the number of nodes to be managed (by a factor of 4–25) and the increase in effort for system configuration and operation across nodes within an overlay/underlay. In the following sections, the various types of heterogeneous-network deployment are discussed in detail.

6.5.2 Indoor distributed-antenna system

Indoor passive distributed-antenna systems (DASs) using coaxial cables are being used by many operators throughout the world to provide LTE coverage in buildings and pavilions prewired with DASs, in high-security areas, and where disruption of equipment maintenance should be kept to a minimum. For an indoor passive DAS, an eNB can be deployed with a DAS (where the radio-frequency connection between the low-power remote radio heads (RRHs) and the baseband processing unit can be over coaxial cables). The advantage of a passive DAS is that signals with nearly uniform powers can be delivered to UEs, which would translate into nearly uniform user data rates across the coverage area. In contrast, if standalone RRHs are deployed, this will result in stronger signals (and higher rates) close to the eNB and weaker signals (and lower rates) at the edge of the coverage area.

An example of an indoor DAS and RRH deployment is shown in Figure 6.12. Next, the performance of the DAS system is simulated for a large hall with approximate dimensions of 200 m by 120 m and compared with an indoor deployment with two RRHs mounted in the ceiling. Two frequency plans are studied. In the first, the same 20-MHz channel is used in both cells (1×20 MHz), whereas in the second, two separate 10-MHz channels (2×10 MHz) are used in a reuse-2 fashion.

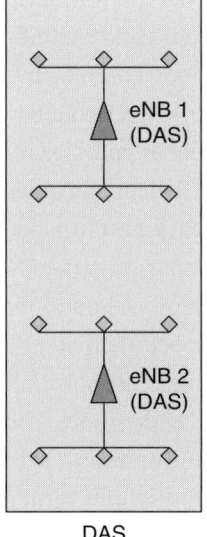
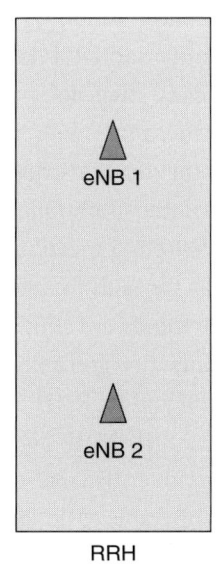

Figure 6.12. Example of indoor DAS and RRH deployment.

DAS   RRH

Table 6.7. *Downlink DAS performance*

Frequency plan	Performance metric (bps/Hz)	DAS	RRH
$1 \times 20\,$MHz	Cell spectral efficiency	2.670	4.321
	Cell-edge UE spectral efficiency	0.065	0.062
$2 \times 10\,$MHz	Cell spectral efficiency	2.685	2.685
	Cell-edge UE spectral efficiency	0.084	0.087

Table 6.8. *Uplink DAS performance*

Frequency plan	Performance metric (bps/Hz)	DAS	RRH
$1 \times 20\,$MHz	Cell spectral efficiency	1.02	1.65
	Cell-edge UE spectral efficiency	0.027	0.033
	IoT (dB)	9.77	10.02
$2 \times 10\,$MHz	Cell spectral efficiency	1.04	1.05
	Fifth-percentile edge UE spectral efficiency	0.025	0.026

Table 6.7 shows the TDD (configuration 1) downlink performance results. It is observed that the traditional RRH deployment yields significantly better cell spectral efficiency than does the indoor DAS with a 1×20-MHz frequency plan. Furthermore, with a 2×10-MHz frequency plan the cell spectral efficiency of the DAS deployment is not significantly different, whereas that of the traditional deployment becomes similar to that of the DAS. The cell-edge spectral-efficiency performance of the two deployments is similar for both frequency plans.

In the uplink, the performance trends are similar to those observed for the downlink: the cell spectral efficiency and cell-edge spectral efficiency with a 1×20-MHz frequency plan and an RRH deployment are substantially better than those for the 2×10-MHz scheme. Furthermore, the values of the interference-power-over-thermal-noise-power ratio (IoT) for the two deployments are comparable. The uplink throughput results are summarized in Table 6.8.

6.5.3 In-band relays

Relays are used to improve the overall sector throughput performance and the coverage of the LTE system. In other words, relays can provide better user experience for UEs situated in poor-geometry locations. Repeaters can also offer enhanced coverage for eNBs by amplifying and forwarding received waveforms, but suffer from the following disadvantages: they can't distinguish between signals and interference/noise, and radio-frequency isolation is very problematic, resulting in interference issues. The LTE system will mainly utilize in-band relays, for which the eNB-to-relay link shares the same band with direct eNB-to-UE links within the cell. Out-of-band relays, for which the eNB-to-relay link does not share the same band with direct eNB-to-UE links, can also be used, but they consume valuable spectrum resources. Relays can either be deployed outdoors or indoors. The use cases for outdoor relays are hotspots and deadspot mitigation. Outdoor relays can be deployed above rooftops, below rooftops, and on street poles. Both omni-directional and directional donor antennas can be used for outdoor relays. The directional donor antenna enhances the in-band backhaul capability, as will be shown later in the chapter. The use of indoor relays is likely to be prevalent in urban as well as rural environments since most of the data traffic is generated indoors. Indoor relays consist of two units: a donor unit, which is generally placed outside the building, and a coverage unit, which is placed inside. The donor and coverage units are connected by a cable or out-of-band wireless link when they are not co-located. The relay donor antenna can be placed on a rooftop so that it has a line of sight to the donor eNB.

The type of relay node which is being standardized in 3GPP is also known as type-1 relays. Type-1 relays have the following characteristics. First, the relay cell has its own physical cell ID, transmits its own synchronization channels, reference symbols, etc., and will be distinct from the donor cell. Second, the UE should receive scheduling information and HARQ feedback directly from the relay node (RN) and send its control channels (SR/CQI/ACK) to the RN. Third, the RN appears as a Rel-8 eNB to Rel-8 UEs. The relay access is split as shown in Figure 6.13.

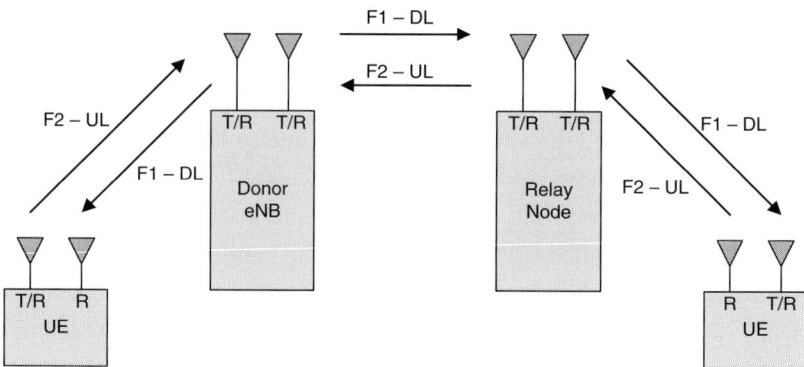

Figure 6.13. Relay access split.

As this figure illustrates, for FDD the eNB-to-RN transmission occurs in the downlink frequency, whereas the RN-to-eNB transmission occurs in the uplink frequency. The maximum number of transmit antennas supported for eNB–RN and RN–eNB links is limited to four.

The design of a robust and efficient in-band backhaul link is essential for efficient in-band relay operation. The backhaul traffic is transmitted by creating gaps in the RN-to-UE transmission in order to avoid simultaneous transmission and reception on the same carrier frequency. Thus the eNB-to-RN and RN-to-UE links are multiplexed in time, and similarly UE-to-RN and RN-to-eNB links are also multiplexed in time. Ideally these gaps can be created by introducing the concept of almost blank subframes (ABSs). However, in the current LTE standard, blank subframes are achieved by configuring certain subframes as MBSFN subframes in the relay cell. It may be noted that MBSFN subframes are not completely blank subframes but contain two control symbols. The concept is illustrated in Figure 6.14, where the utilization of subframes at eNBs, RNs, macro-cell UEs (UE1), and relay-cell UEs (UE2) is shown. The arrows show the direction of transmission for radio links in each subframe. Thus, subframes are normally used for access links, i.e. downlink transmission from an eNB or RN to its UEs, except during the MBSFN subframes, when UEs in the relay cell do not receive data, whereas eNBs may transmit downlink traffic both to RNs (i.e. backhaul

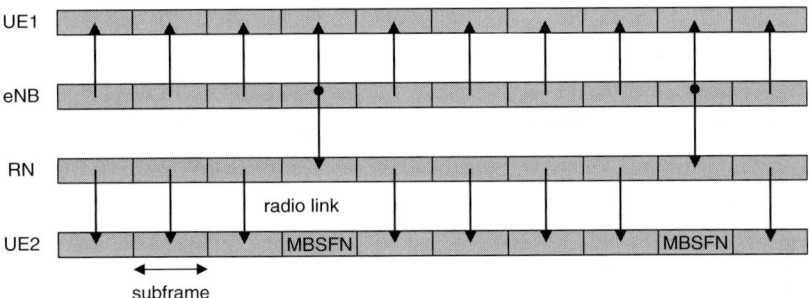

Figure 6.14. Transmission of backhaul traffic by creating gaps in RN to UE transmission.

traffic) and to macro-cell UEs (access traffic). The set of ABSs that can be configured for downlink eNB–RN transmission and uplink RN–eNB transmission for both frame structure type 1 and frame structure type 2 is given in [21]. The set of such downlink backhaul subframes is semi-statically assigned, whereas the set of uplink backhaul subframes is either semi-statically assigned or implicitly derived from the downlink backhaul subframes using the HARQ timing relationship.

A new physical control channel, the Relay-PDCCH (R-PDCCH), is defined for the eNB to assign backhaul resources to the RN. One requirement for the R-PDCCH is the increased flexibility in multiplexing (eNB-to-RN and eNB-to-UE traffic). For example, when the RN serves no UEs or has no backhaul traffic, then it should be possible to assign the resource blocks used for the R-PDCCH to the Rel-8 UEs. The R-PDCCH also facilitates sending backhaul control information for an LTE TDD system without which the operation of the TDD system would have been complex. The R-PDCCH has the following characteristics. First, it supports both common reference symbols and dedicated reference symbols. Second, the downlink scheduling assignments are always transmitted in the first slot of a subframe. If a scheduling assignment is transmitted in the first PRB of a given PRB pair, then an uplink scheduling assignment may be transmitted in the second PRB of the PRB pair. Third, the R-PDCCH is multiplexed in a TDM+FDM fashion. Finally, the R-PHICH is not supported and HARQ feedback is included in the

R-PDCCH. The R-PDCCH transmission format can be either without cross-interleaving or with cross-interleaving. In the cross-interleaved mode, the R-PDCCHs for different relay nodes are multiplexed within the same PRB, whereas for the non-interleaved case the R-PDCCHs for different relay nodes are not multiplexed within the same PRB. It may be noted that the search-space design for detecting R-PDCCHs is different for these two formats.

For RN–eNB transmission, the PUCCH is used to send the HARQ acknowledgments corresponding to decoding of the PDSCH and the scheduling request message is transmitted only in the uplink subframes that have been configured for RN–eNB transmission.

Next, the performance of outdoor relays with in-band backhaul is shown and compared with that of relays with out-of-band backhaul. Four models are considered for the in-band backhaul link when simulating outdoor relays, as described below. The models to be used for pathloss, antennas, and lognormal shadowing on the access and backhaul links are described in [22].

1. Backhaul A: non-optimized relay-site planning with a single, omni-directional antenna set at the RN.
2. Backhaul B: the backhaul model for optimized relay-site planning (bonus pathloss of 5 dB) with a single, omni-directional antenna set.
3. Backhaul C: backhaul A with a directional antenna for the backhaul link and an omni-directional antenna set for the relay-access links.
4. Backhaul D: backhaul B with a directional antenna for the backhaul link and an omni-directional antenna set for the relay-access links.

A two-ring, 19-macro-cell, three-sectored site hexagonal grid system layout is simulated with dual-port UE receiver operation and assuming TU channels using cell wrap-around for two systems, each operating in a 10-MHz bandwidth, corresponding to a deployment scenario with inter-site distance 1.732 km. In this simulation 1425 UEs are randomly dropped with uniform spatial probability density over the entire 57-cell network. For the results presented here, a deployment with 228 relays is considered. The relays are dropped randomly over the entire network with a uniform spatial distribution.

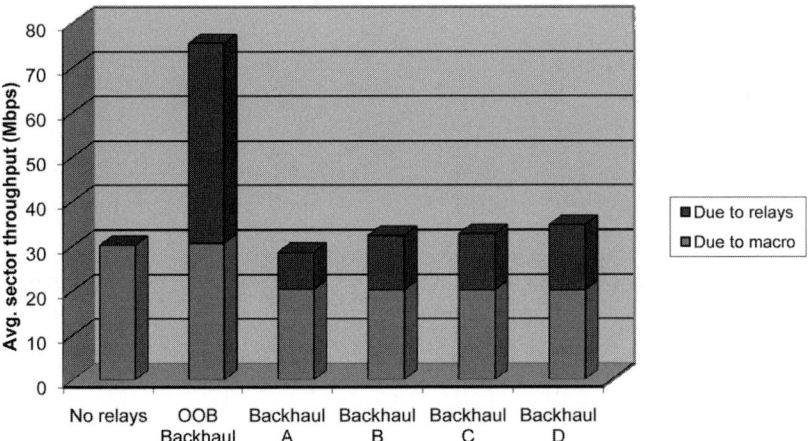

Figure 6.15. Average sector throughput performance of macro-cell plus outdoor relays.

By controlling the number of backhaul subframes per frame, the size of the backhaul pipe can be controlled (at the expense of the resources available for the macro-cell access links). Note that this is the total number of subframes shared by all backhaul links in the sector. For in-band backhaul, the scheduler of each RN is constrained to allocate resources to its UEs only when the amount of data that it has transferred to the UEs does not exceed the amount of data the RN has received from the donor eNB. This constraint ensures that the relay-cell throughput does not exceed the corresponding backhaul throughput. For out-of-band (OOB) backhaul simulations, it is assumed that the backhaul is ideal and unconstrained.

In Figure 6.15, the aggregate sector throughput is shown as the sum of its two components: that due to the macro-cell and that due to all RNs (when present) associated with a donor macro-cell. Clearly, ideal OOB backhaul yields the best throughput performance since unlimited backhaul capacity is assumed to be freely available. With in-band backhaul, backhaul D performs best, being aided by both optimized relay-site planning and a directional antenna. On the basis of the simulation results, in-band relays will mainly be used to fill coverage holes where wired or OOB wireless backhaul will be difficult to deploy.

6.5.4 Pico- and femto-cell underlay

The LTE pico or femto system is a low-power access solution that is intended for deployment as an underlay in a macro network. Its primary purpose is to provide enhanced capacity for busy outdoor areas and improved coverage for indoor areas for an existing macro network. The deployment of low-power pico nodes below the rooftop will allow more street-level coverage and increase the spectrum efficiency of the overall system. Simulation results for low-power pico nodes deployed within a macro-cell both for the downlink and for the uplink are shown in this section. The simulations are performed as per [22]. The key simulation assumptions can be summarized as follows. A 19-cell, three-sector macro-cell system with wrap-around is defined and 2, 4, and 10 pico-cells are randomly dropped within each macro-cell coverage area. The macro-cell transmission power is 40 W while the pico-cell transmission power is 1 W. Users attach either to the macro-cell or to the pico-cell depending on the strength of the received reference-signal power measurement. Owing to the large power difference, cell attachment is biased in favor of the macro-cell. For instance, with one macro-cell and one pico-cell in the coverage area, 92% of the users will attach to the macro-cell.

Overall downlink performance results when pico-cells are added into the macro-cell coverage area are shown in Table 6.9. Both user and sector throughput results within the macro-cell coverage area are shown. For the user throughput, the 50th-percentile results are

Table 6.9. *Downlink performance with pico-cells*

Number of pico-cells per macro-cell	50th-Percentile user throughput (bps)		Total sector throughput	Fraction of UEs associated with pico-cells (%)
	Macro-cell	Pico-cell		
0	5.0×10^5	–	1.6×10^7	–
1	5.3×10^5	3.2×10^6	2.5×10^7	7.9
2	5.8×10^5	3.6×10^6	3.4×10^7	14.7
4	6.7×10^5	3.8×10^6	5.2×10^7	26.3
10	9.4×10^5	5.0×10^6	9.5×10^7	44.4

shown. The percentages of users associated with pico-cells are also shown in Table 6.9. It can be observed that, due to the large difference in transmission power and antenna gain between macro-cell and pico-cell eNBs, even UEs dropped in the vicinity of a pico-cell (within 40 m of the pico-cell) can be associated with a macro-cell instead of the pico-cell. However, user throughput over the cell area in a pico-cell is approximately six times better than that in a macro-cell. It may be noted that the macro and pico nodes use the same carrier frequency and that no interference-mitigation techniques are used. As the number of pico nodes per sector increases, the macro-cell user experience improves since only the good-geometry users are connected to macro-cells and more traffic is offloaded to the pico-cells. For example, with one pico-cell per sector, 2 UEs are served by the pico-cell and 28 by the macro-cell. With four pico-cells per sector, 8 UEs are served by the pico-cells and 22 by the macro-cell. So the macro-cell UE performance improves because there are fewer UEs to serve. The overall UE throughput improves because more pico-cells are present.

Overall uplink performance results when pico-cells are added into the macro-cell coverage area are shown in Table 6.10. From the results, it can be seen that uplink user performance is significantly improved with the

Table 6.10. *Uplink performance with pico-cells*

Number of pico-cells per macro-cell	50th-Percentile user throughput (bps)		Total sector throughput	Fraction of UEs associated with pico-cells (%)	IoT (dB)
	Macro-cell	Pico-cell			
0	2.0×10^5	–	8.8×10^6	–	9.7
1	2.1×10^5	4.31×10^6	1.9×10^7	7.9	9.9
2	2.3×10^5	4.12×10^6	2.8×10^7	14.7	10.3
4	2.7×10^5	4.01×10^6	4.3×10^7	26.3	10.4
10	3.6×10^5	4.05×10^6	7.3×10^7	44.4	10.5

addition of pico-cells. For instance, without pico-cells, the median user throughput is 200 kbps. With one pico-cell added, this improves to 210 kbps. With 10 pico-cells added, the average user throughput increases to 360 kbps. The improvement increases with the number of pico-cells per sector as traffic is offloaded onto the pico-cells. The total sector throughput is also seen to increase significantly with the addition of pico-cells. For instance, the overall throughput improves by 117% when 1 pico-cell is added, and by 725% when 10 pico-cells are added into the macro-cell coverage area.

Some system statistics are also presented in Table 6.10. It can be seen that the fraction of UEs associated with the pico-cells is 15%, 26%, and 44%, respectively, for 2, 4, and 10 pico-cells per macro-cell. Although there is a large difference in transmission power and antenna gain between macro-cell and pico-cell eNBs, in general an acceptable percentage of UEs is attached to the pico-cells. Note that biased cell selection is not used in this study. Such a technique can be used to further increase the association percentage if so desired. Also note that the IoT is kept to approximately 10 dB using appropriate power-control settings regardless of the deployment scenario. In this case, it is seen that the rise in uplink IoT is small even when more UEs are transmitting simultaneously.

6.5.5 Interference-management techniques for heterogeneous network

The results presented in the previous section were achieved without using any ICIC techniques. Since the interference scenario in heterogeneous network deployments is different from that in macro-cell-only deployments, effective ICIC techniques will further enhance the performance of the heterogeneous network system. An example of the interference scenario is depicted in Figure 6.16, wherein a UE connected to the macro-cell may experience interference from the femto-cells or the pico-cells on the downlink and may also cause interference with heterogeneous network nodes in the uplink direction. If a UE that is not part of the femto-cell closed subscriber group (CSG) is in the vicinity of the pico-cell coverage, it will experience interference in the downlink

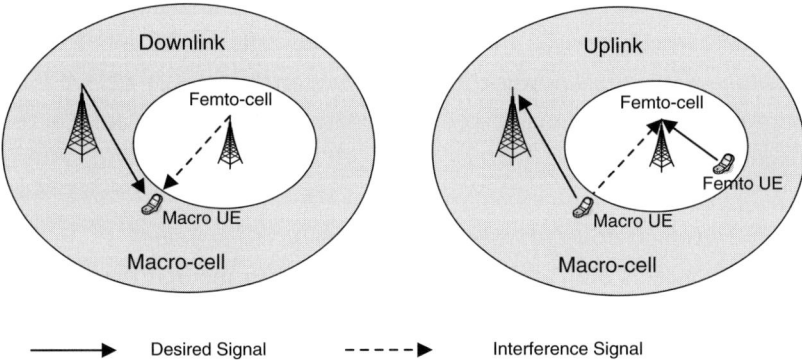

Figure 6.16. Heterogeneous network interference scenarios.

direction from the femto-cell downlink. In the uplink direction, the non-CSG UE will be connected to a macro-cell and will be transmitting at close to full power, thus creating interference with the femto-cell uplink. In another example, if a macro-cell wants to offload its users to pico-cells, a cell-range-extension technique (e.g. biasing) is required, which will require interference mitigation since the users might not be connected to the best serving cell. Hence, ICIC techniques will be beneficial for the effective performance of the heterogeneous network underlaid in a macro-cell network.

The ICIC techniques for heterogeneous networks can be classified under three main categories, namely Rel-8/9-based schemes, Rel-10-based schemes without carrier aggregation, and Rel-10 schemes based on carrier aggregation. The ICIC techniques generally require synchronization between the macro and heterogeneous network nodes.

6.5.5.1 Rel-8/9-based schemes

One of the simplest ICIC schemes involves using different carrier frequencies for different cell layers. As an example, carrier frequencies f_1 and f_2 are used for the overlay macro-cell, with carrier f_2 being transmitted at a lower power from the macro-cell and carrier f_2 being used in the underlay pico-cell. Standard Rel-8/9 semi-static ICIC schemes whereby RNTP can be exchanged between a pico node and a macro-cell over the X2 interface can also be used. Preliminary simulation results show that the user experience can be improved significantly when two separate carriers are used for macro- and pico-cells. In the uplink, optimization of fractional power-control schemes on the basis of X2 overload control can mitigate the inter-cell interference and the rise in noise at the macro- and pico-cells.

6.5.5.2 Rel-10 non-carrier-aggregation-based schemes

Several concepts are currently being discussed in the 3GPP standards. These are based on both frequency- and time-domain techniques and are applicable both for the data channels and for the control channels. In the time-domain scheme for the PDSCH and PDCCH, the data from the macro-cell is transmitted, for example, on alternate subframes. The gaps in downlink data transmission are created using almost blank subframes (i.e. MBSFN subframes) in the macro-cell. Data is transmitted from the pico-cell during all the subframes. In other words, the low-power pico-cell uses all its resources while the macro-cell uses a fraction of its resources. A bitmap is used to indicate the almost blank subframe pattern of the macro-cell, which is transmitted to pico-cells using X2 signaling. The pattern period can be between 40 ms and 70 ms (using FDD or TDD) and is semi-statically configured. Additionally, the RRC signaling is modified to provide resource-specific RLM/RRM measurements both for the serving cell and for the neighbor cell.

Additionally, Rel-8 interference-management techniques using backhaul can be used to further improve the performance of the data channel. In the frequency-domain scheme, part of the time-frequency resources may be reserved in the macro-cell, where the PDSCH is not transmitted while the PDSCH is transmitted over the whole time-frequency resource in the pico-cell. For the control channel, different parts of the carrier bandwidth can be used to transmit the PDCCH in different cell layers. Both these techniques extend the coverage of the low-power node and improve the reliability of the control and data transmission.

6.5.5.3 Rel-10 carrier-aggregation-based schemes

For Rel-10 UEs connected to the macro-cell and pico-cell the control can be sent on carrier f_1 and carrier f_2, respectively, so that there is no

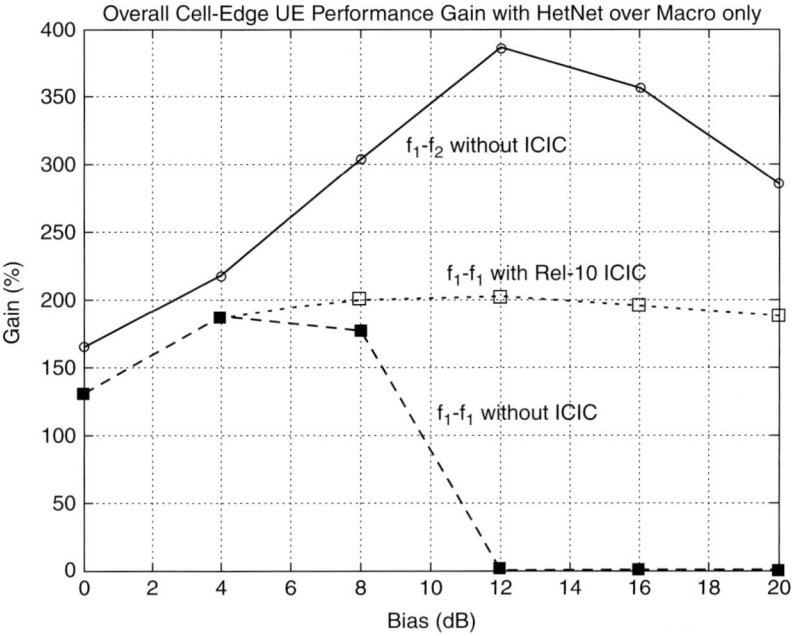

Figure 6.17. Performance of Rel-10 ICIC scheme with one and two carriers.

control-channel interference. On the other hand, data for Rel-10 UEs can be scheduled on multiple component carriers (in this case f_1 and/or f_2) using carrier aggregation with cross-carrier scheduling. Using carrier-aggregation techniques, each grant contains a carrier-indication field (CIF) to indicate which carrier the grant applies to, thus allowing reuse of existing LTE scheduling grant formats with only the addition of the CIF. Additionally, for the data part, downlink interference-coordination techniques can be used.

The cell-edge and average UE throughput can be significantly improved by modifying the cell-attachment procedure. By biasing cell selection, the UEs in the vicinity of pico-cells can be attached to pico-cells, resulting in traffic offload from the macro-cells to the pico-cells. This in turn maximizes the performance gain provided by the addition of these pico-cells to the network. Range expansion is typically character-ized by associations where the serving cell does not correspond to the cell with the best downlink geometry for a given user; instead, the serving cell

Table 6.11. *Additional user equipment category for LTE Rel-10*

UE category	Maximum number of bits in a subframe		Maximum number of downlink MIMO layers
	Downlink	Uplink	
6	301 504	51 024	2 or 4
7	301 504	102 048	2 or 4
8	2 998 560	1 497 760	8

corresponds to, for example, the cell with lowest pathloss to the UE. Figure 6.17 shows (i) improvement in overall edge-user throughput with an almost blank subframe as the bias is increased and (ii) improvement in overall edge-user throughput when the macro and pico overlays are supported on two different carriers.

6.6 Miscellaneous

In addition to the major enhancements described earlier, several other miscellaneous physical-layer enhancements have also been introduced in LTE-A. They include non-contiguous uplink transmission and aperiodic sounding reference signals. Further, additional UE categories as shown in Table 6.11 were defined in order to support higher throughputs in LTE-A.

6.6.1 Non-contiguous uplink transmission

In LTE, uplink data allocation must be contiguous in order to preserve a low cubic metric. Two drawbacks of this requirement are limited frequency-selective scheduling gain and resource fragmentation. The gain from frequency-selective scheduling is limited because the preferred resource blocks cannot be assigned to the user unless they are contiguous. In addition, the frequency resource can be fragmented because retransmissions generally occupy the same bandwidth as initial transmissions. Thus, these retransmissions leave holes in the frequency blocks that might not be completely filled, leading to less than 100% resource utilization. These two drawbacks can be addressed with non-contiguous uplink

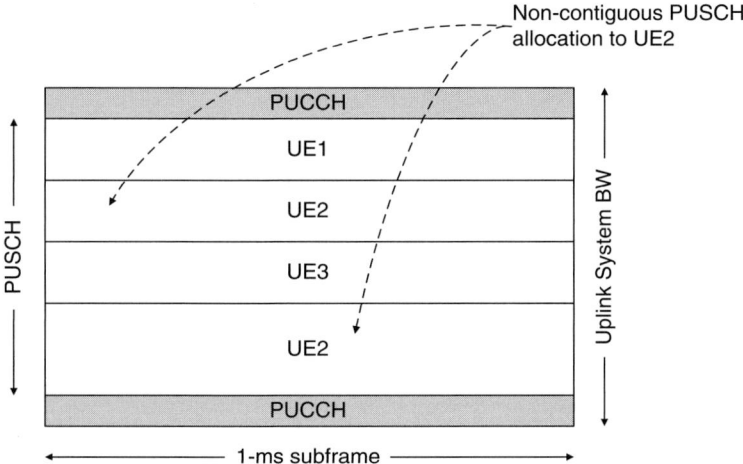

Figure 6.18. Example of non-contiguous uplink data transmission.

transmission, where a user is allowed to transmit on two non-contiguous blocks simultaneously. However, with the introduction of non-contiguous uplink transmission, the UE and system complexity will increase. In addition, the loss in terms of the cubic metric will negate some of the system gain.

In LTE-A, UE can be scheduled to transmit two PUSCH clusters simultaneously as illustrated in Figure 6.18. The main advantage of non-contiguous resource allocation within one component carrier is the frequency-selective scheduling gain. In [23], gains from non-contiguous resource allocation were analyzed. From the results shown, it can be seen that there is a gain of approximately 12%–15% in sector throughput and a gain of 17%–25% in cell-edge throughput. However, in these analyses, ideal channel knowledge and channel estimation were usually assumed. For realistic scenarios, the sector throughput gain reduces to approximately 4%–5%.

In addition, with non-contiguous resource allocation, PUSCH resource fragmentation can be eliminated or greatly reduced. Resource fragmentation occurs when certain resource blocks cannot be assigned due to the single-carrier requirement. As a result, a loss of spectral efficiency can occur. In [24], a 6.5% improvement in bandwidth utilization was observed when non-contiguous allocation was used. In general, this problem becomes less severe as the number of available users in the system increases, since more users can be scheduled to fill in the resource gap. Thus, in a typical deployment with a large number of users in a cell, the loss of spectral efficiency is expected to be very small. However, it should be noted that, even with a sufficient number of users, control-channel limitations may prevent a large number of users being scheduled in one subframe. Thus, it may still be beneficial to use non-contiguous allocation. With non-contiguous resource allocation, the issue of PUSCH resource holes can be greatly reduced or even eliminated while keeping the control overhead small.

6.6.2 Aperiodic SRS

Dynamic aperiodic SRS will be supported in Rel-10 in order to increase SRS capacity and provide sounding capability as needed by the eNB. Similarly to aperiodic CQI in LTE Rel-8, users will be triggered to transmit aperiodic SRSs. This will be done using control signaling via an uplink scheduling grant. In this case, the underlying assumption is that SRS triggering occurs at the same time as a PUSCH allocation. This makes sense insofar as SRS can be used to support uplink data-transmission techniques such as frequency-selective scheduling and uplink MIMO. However, SRSs may be transmitted on a different sub-frame than data transmission. In this case, an SRS-request bit similar to the aperiodic CQI request bit will be added. When triggered, a user will transmit an SRS using a predefined configuration. Note that a user may be configured for both periodic and aperiodic SRS transmission. For instance, periodic SRS transmission with a very long period may be configured for the UE in order to provide long-term channel statistics. When uplink data is available, the eNB can then trigger the UE to transmit an aperiodic SRS prior to uplink data scheduling.

References

[1] 3GPP TS 36.913, Requirements for further advancements for Evolved Universal Terrestrial Radio Access (E-UTRA) – (LTE-Advanced), v9.0.0, December 2009.

[2] Ghosh, A., Ratasuk, R., Mondal, B., Mangalvedhe, N., Thomas, T., "LTE-advanced: next-generation wireless broadband technology," *IEEE Wireless Communications*, vol. 17, no. 3, pp. 10–22, June 2010.

[3] Osseiran, A., Hardouin, E., Gouraud, A. *et al.*, "The road to IMT-advanced communication systems: state-of-the-art and innovation areas addressed by the WINNER + project," *IEEE Communications Magazine*, vol. 47, no. 6, pp. 38–47, June 2009.

[4] Iwamura, M., Etemad, K., Mo-Han, F., Nory, R., Love, R., "Carrier aggregation framework in 3GPP LTE-advanced," *IEEE Communications Magazine*, vol. 48, no. 8, pp. 60–67, August 2010.

[5] R1-082468, "Carrier aggregation in LTE-Advanced," Ericsson, RAN1#53bis, Warsaw, June 2008.

[6] 3GPP TS 36.104, Base station (BS) radio transmission and reception, v9.1.0, September 2009.

[7] R4-090963, "Prioritized deployment scenarios for LTE-Advanced studies," NTT DoCoMo *et al.*, RAN4#50, Athens, February 2009.

[8] 3GPP TS 36.942, Radio frequency (RF) system scenarios, v8.2.0, May 2009.

[9] R4-091749, "Co-existence studies of contiguous aggregation deployment scenarios for LTE-A", Motorola, RAN4#51, San Francisco, CA, May 2009.

[10] R1-101467, "PHICH for carrier aggregation," Motorola, RAN1#60, San Francisco, CA, February 2010.

[11] R1-105096, "36.211 Draft CR (Rel-10, B) Introduction of Rel-10 LTE-Advanced features," Ericsson, RAN1#62, Madrid, August 2010.

[12] R1-090795, "UL-MIMO with antenna gain imbalance," Motorola, RAN1#56, Athens, February 2009.

[13] R1-093327, "Tx diversity for LTE-Advanced PUSCH," Nokia Siemens Networks, Nokia, RAN1#58, Shenzhen, China, August 2009.

[14] R1-100506, "Further investigation on necessity of PUSCH Tx diversity for LTE-Advanced," NTT DoCoMo, NEC Group, RAN1#59bis, Valencia, January 2010.

[15] R1-090865, "CoMP cooperative silencing hotzone DL performance," Qualcomm Europe, RAN1#56, Athens, February 2009.

[16] Li, Q., Fang, S., Yang, Y., Pan, Z., "User pairing transmission scheme in uplink coordinated multi-point reception," *2nd International Conference on Future Computer and Communication*, May 2010.

[17] Venkatesan, S., "Coordinating base stations for greater uplink spectral efficiency in a cellular network," *IEEE 18th International Symposium on Personal, Indoor and Mobile Radio Communications*, September 2007.

[18] Falconetti, L., Hoymann, C., Gupta, R., "Distributed uplink macro diversity for cooperating base stations," *IEEE International Conference on Communications Workshops*, June 2009.

[19] R1-090823, "Discussion on timing advance issue in CoMP & text proposal," Huawei, RITT, Texas Instruments, CMCC, RAN1#56, Athens, February 2009.

[20] Cisco, "Cisco visual networking index: forecast and methodology, 2009–2013," June, 2010.

[21] 3GPP TS 36.216, "Physical layer for relaying operation," v10.0.0, October 2010.

[22] 3GPP TS 36.814, "Further advancements for E-UTRA physical layer aspects," v9.0.0, March 2010.

[23] R1-084582, "System performance of uplink non-contiguous resource allocation," Panasonic, RAN1#55, Prague, November 2008.

[24] R1-084398, "Analyses comparing different UL transmission schemes for LTE-A," Qualcomm Europe, RAN1#55, Prague, November 2008.

Additional reading

[1] Loa, K., Chih-Chiang, W., Shiann-Tsong, S. *et al.*, "IMT-advanced relay standards," *IEEE Communications Magazine*, vol. 48, no. 8, pp. 40–48, August 2010.

[2] Yu, C., Hua, C., "Cooperative broadcasting schemes for LTE-A," *2009 IEEE 20th International Symposium on Personal, Indoor and Mobile Radio Communications*, pp. 1487–1491, September 2009.

[3] Minghai, F., Xiaoming, S., Lan, C., Kishiyama, Y., "Enhanced dynamic cell selection with muting scheme for DL CoMP in LTE-A," *IEEE 71st Vehicular Technology Conference*, May 2010.

[4] Young-Han, N., Lingjia, L., Yan, W. *et al.*, "Cooperative communication technologies for LTE-advanced," *IEEE International Conference on Acoustics Speech and Signal Processing (ICASSP)*, pp. 5610–5613, March 2010.

[5] Xiaolin, H., Zhan, Z., Kayama, H., "Adaptive multi-Tx multi-Rx MIMO transmission scheme for LTE-Advanced downlink," *IEEE Global Telecommunications Conference*, December 2009.

[6] Jialing, L., Love, R., Nimbalker, A., "Recent results on relaying for LTE-Advanced," *IEEE 70th Vehicular Technology Conference*, September 2009.

[7] Sawahashi, M., Kishiyama, Y., Morimoto, A., Nishikawa, D., Tanno, M., "Coordinated multi-point transmission/reception techniques for LTE-advanced," *IEEE Wireless Communications*, vol. 17, no. 3, pp. 26–34, June 2010.

7 Comparison of broadband technologies

7.1 Introduction

LTE Rel-8 and WiMAX are the two main wireless broadband technologies based on OFDM which are currently being commercialized. Both of these technologies are being enhanced (LTE-Advanced and 802.16m) so as to support higher peak rates, higher throughput and coverage, and lower latencies, resulting in a better user experience. Further, both LTE-Advanced and 802.16m were approved by the ITU as IMT-Advanced technology. Also several operators are considering deploying both these technologies or migrating their existing WiMAX system to LTE or 802.16m. In this chapter, these two main broadband technologies are compared with respect to their features and system performance. Also, WiMAX and LTE co-existence and migration scenarios are briefly discussed.

7.2 Feature comparison of wireless broadband technologies

The primary competing wireless broadband technology to 3GPP LTE is a family of standards developed by the Institute of Electrical and Electronic Engineers (IEEE) called 802.16. WiMAX is an industry forum promoting 802.16 that also defines a subset of features (called profiles) based on 802.16 specifications. Since the IEEE 802.16 family of standards contains many optional features that need not be implemented by vendors and operators, the task of the WiMAX forum is to decide on an agreed profile by reducing the number of options in the 802.16 specifications and to promote inter-operability among equipment vendors and system operators. The 802.16-2004 standard (known as 802.16d) uses OFDM access both for the downlink and for the uplink, and is used for fixed applications. Scalable bandwidth operation (i.e. with the symbol duration and inter-subcarrier spacing constant irrespective of the bandwidth) and mobility enhancements were provided in the next-generation 802.16e standards. The 802.16e standard was first published in early 2006, two years prior to LTE Rel-8. The WiMAX Rel-1 profile supports TDD transmission, various frequency-permutation schemes, and multi-antenna technology. Although there are quite a lot of similarities in features, there are some major differences between the WiMAX Rel-1 profile and LTE Rel-8. The primary physical-layer differences between these two technologies are as follows.

Uplink transmission schemes. LTE uses single-carrier FDMA (SC-FDMA), whereas OFDMA is used for WiMAX. As discussed in Section 2.3.2, SC-FDMA requires lower amplifier power back-off, thus improving the cell-edge performance of the system compared with OFDMA. For the same maximum transmit power, an LTE user is capable of transmitting using approximately 2 dB higher power on the cell edge due to the use of SC-FDMA. This results in LTE providing either greater cell coverage or a higher data rate for the same coverage. However, the link performance of SC-FDMA is inferior to that of an OFDMA receiver, especially for higher-order modulations. Thus, cell-center users may experience slight performance loss under LTE. Finally, the control-channel design of an SC-FDMA system has to become somewhat complex in order to maintain the single-carrier property of the system. In Rel-10, carrier aggregation is supported and the single-carrier property of SC-FDMA is no longer preserved, so there is a reduction in coverage when carrier aggregation is used. In 802.16m, since the transmission scheme is based on OFDMA, there is no effect on coverage when carriers are aggregated.

Downlink control-channel design. A requirement for the downlink control-channel design is that the control information should be reliably decoded at the cell edge without the aid of HARQ. As indicated in Section 3.5, there are two principles for control-channel design, namely a broadcast common control channel and a per-user dedicated control channel. In the case of a broadcast control channel, the control signals for multiple users are jointly coded. The code rate is designed to decode the worst-geometry user, and therefore the control signals cannot be individually power-controlled or beamformed. In WiMAX Rel-1, the control-channel design is based on a broadcast control channel, whereas in LTE it is based on a per-user control-channel design. It may also be noted that in a WiMAX system coverage is limited by the control channel in the downlink and by the data channel in the uplink. In 802.16m, the control-channel design was modified to accommodate a per-user control channel. A new MAP called Advanced-MAP that carries resource-allocation information both for the downlink and for the uplink every subframe was defined.

Multi-antenna support. IEEE 802.16e supports a host of multiple-antenna-system (MAS) technologies, but the downlink multi-antenna technologies supported by the WiMAX profile can be generally classified into open-loop MIMO schemes that include space–time block coding (STBC or MIMO-A), open-loop spatial multiplexing (MIMO-B) with adaptive mode selection between the two, and UE-specific reference-symbol-based beamforming. The transmitter using open-loop MIMO does not require explicit knowledge of the fading channel. The downlink multi-antenna transmission modes for LTE Rel-8 have been discussed in Section 5.2. In LTE Rel-8 up to five principal multiple-antenna modes are supported in the downlink, namely transmit diversity, open-loop spatial multiplexing (OLSM), closed-loop spatial multiplexing (CLSM or single-user MIMO (SU-MIMO)), multi-user MIMO (MU-MIMO), and UE-specific reference-symbol-based beamforming. Both WiMAX and LTE can support up to eight transmit antennas at the base station and four receive antennas at the UE. In the uplink, both WiMAX and LTE Rel-8 support one transmit and up to eight receive antennas, and the only multi-antenna scheme supported on the uplink for both of these technologies is MU-MIMO. In LTE Rel-10, the downlink spatial multiplexing scheme was extended to support 8×8 MIMO and enhanced MU-MIMO based on dedicated reference symbols. Similar enhancements have been made to 802.16m. Table 7.1 shows the similarities/differences in MAS schemes between LTE Rel-10 and 802.16m. In LTE Rel-10 and 802.16m, uplink spatial multiplexing supporting up to four streams is introduced. Hence the peak data rate supported on the uplink is quadrupled. Also, transmit diversity is supported for the LTE Rel-10 control channel.

Frame structure. There is a difference in transmission parameters and frame structure between WiMAX and LTE. With respect to transmission parameters, the subcarrier spacings for WiMAX and LTE are 11 and 15 kHz, respectively, which difference does not have a major impact on

Table 7.1. *Comparison of MIMO modes for WiMAX and LTE*

MIMO technology	IEEE 802.16e (WiMAX R1)	IEEE 802.16m (WiMAX R2)	LTE-Rel-8	LTE-Rel-10
SISO/SIMO	Single transmit zone (SISO/SIMO or CSTD)	No explicit SISO/SIMO mode (minimum number of BS transmit antennas is 2)	Transmit mode 1	Transmit mode 1
Open-loop transmit diversity	Matrix A (STC zone)	MIMO mode 0 (SFBC with RP), MIMO mode 5 (CDR)	Transmit mode 2 (SFBC with FSTD)	Transmit mode 2 (SFBC with FSTD)
Open-loop SU-MIMO	Matrix B (STC zone)	MIMO mode 1 (rank 1 & SM with RP)	Transmit mode 3 (SM with large-delay CDD)	Transmit mode 3 (SM with large-delay CDD)
Closed-loop SU-MIMO (codebook-based)	Not in profile	MIMO mode 2 (rank 1 & SM with AP)	Transmit modes 4 & 6 (rank 1 & SM)	Transmit mode 9 [up to rank 8 & SM]
Closed-loop SU-MIMO (UL sounding-based)	MIMO (A & B) + BF (STC zone with dedicated pilots)	MIMO mode 2 (Rank 1 & SM with AP)	Transmit mode 7 (UE-specific BF (rank 1 only))	Transmit modes 8 & 9 (up to rank 2 for mode 8 and rank 8 for mode 9)
Closed-loop MU-MIMO	Not in profile	MIMO mode 4 (codebook & sounding-based)	Transmit mode 5 (codebook-based only)	Transmit modes 8 & 9
Open-loop MU-MIMO	No equivalent version	MIMO mode 3 (fixed precoding)	No equivalent version	No equivalent version

performance. The TDD frame size in WiMAX is 5 ms while the subframe size in LTE is 1 ms, thus allowing LTE to have lower user-plane latency and lower HARQ turnaround time. Finally, both WiMAX and LTE support cyclic-prefix lengths that are comparable and have been designed to combat multipath fading in different environments. The subframe structure in 802.16m is similar to that in LTE. There are eight subframes of size 0.617 ms per 802.16e frame. Hence the user-plane latency of 802.16m is comparable to that of LTE.

Downlink and uplink sub-channelization. Both WiMAX and LTE support various sub-channelization schemes. In the downlink, the sub-channelization schemes for both these technologies are based on distributed or localized allocation. In WiMAX terminology, distributed transmission is known as partial usage of sub-channels (PUSC), whereas localized transmission is known as adaptive modulation and coding (AMC) mode. In the LTE Rel-8 uplink, both localized and distributed (via PUSCH hopping) allocations are supported. Similarly, the uplink of WiMAX supports both PUSC and AMC sub-channelization schemes.

Downlink and uplink pilot structure. The downlink pilot structure in WiMAX is based on frequency-division multiplexing, and supports both common and dedicated reference symbols, while the uplink pilots are based on a frequency-division-multiplexing/time-division-multiplexing structure. The Rel-8 LTE also supports both common and dedicated reference signals, and the pilot structure is based on frequency-division multiplexing/time-division multiplexing both for the downlink and for the uplink. In LTE Rel-10, the dedicated reference signal is extended to support up to eight streams.

Uplink power control. WiMAX uses open-loop power control with closed-loop corrections on the uplink, whereas LTE Rel-8 uses fractional intra-cell plus X2-based inter-cell power control with closed-loop corrections. The LTE fractional power control is described in Section 4.8. The sector and edge throughput performance of an LTE system as well as the interference rise over thermal are improved significantly using uplink fractional power control.

Hybrid ARQ (HARQ). Both WiMAX and LTE support an N-channel HARQ stop-and-wait protocol, but the total HARQ processing time differs for these two technologies. In LTE FDD Rel-8 and WiMAX Rel-1, the values $N = 8$ and $N = 3$ result in total HARQ processing times of 8 ms and 15 ms, respectively. For LTE TDD, the HARQ processing time is dependent on the downlink/uplink split, and is as shown in Section 4.3.4. Also LTE and 802.16m support HARQ on the basis of a Chase and incremental-redundancy combining scheme, whereas WiMAX profile 1 supports Chase combining only.

Modulation and channel coding. Both LTE Rel-8 and WiMAX Rel-1 support QPSK, 16-QAM, and 64-QAM modulation on the downlink, and QPSK and 16-QAM modulation on the uplink. In addition, LTE Rel-8 supports 64-QAM modulation on the uplink for the highest UE category. Both these technologies support some variant of turbo code, convolutional code, and block codes. However, the granularity of modulation and coding levels supported in LTE is much higher than that in WiMAX Rel-1.

Overhead. Because of the broadcast nature of the control channel in WiMAX, the total downlink overhead varies between 43% and 50% and is inversely proportional to the bandwidth. For LTE, the total overhead is on the order of 30%–32%. The downlink overhead in LTE-A and 802.16m can be reduced to below 30% by using UE-specific reference signals and by reducing the set of common reference signals.

A feature comparison between these technologies is summarized in Table 7.2. It may be observed from this table that similar features are supported for both LTE and 802.16m.

Finally, both LTE-A and 802.16m are designed to be fully backward compatible with LTE and WiMAX Rel-1, respectively. Since the frame size in 802.16m was made smaller than that in WiMAX Rel-1, the WiMAX and 802.16m frames are TDM multiplexed in the downlink and TDM/FDM multiplexed in the uplink.

7.3 Performance comparison of LTE/LTE-A and WiMAX/ 802.16m

The peak data rates supported by WiMAX and LTE are summarized in Tables 7.3 and 7.4, respectively. Although the peak rates do not have true values with respect to system performance, they are of significant value

Table 7.2. Comparison of WiMAX and LTE

Feature	IEEE 802.16e (WiMAX R1)	IEEE 802.16m (WiMAX R2)	LTE Rel-8	LTE Rel-10
Network architecture	Flat, IP-based, BS + ASN GW	Flat, IP-based, BS + ASN GW	Very-flat, IP-based eNB + S-GW	Same as for LTE Rel-8. For heterogeneous network architecture may be different
Mobility	Mobile IP with targeted mobility < 120 km/h	Mobile IP with targeted mobility < 120 km/h	Full 3GPP mobility with target up to 500 km/h, 2G/3G handover with global roaming	Full 3GPP mobility with target up to 500 km/h, 2G/3G handover and global roaming
Access technology	Scalable OFDMA in uplink & downlink	Scalable OFDMA in uplink & downlink	Downlink: scalable OFDMA Uplink: scalable SC-FDMA	Same as for LTE Rel-8. Single-carrier property is not preserved for SC-FDMA uplink
Channel bandwidth	3.5, 5, 7, 8.75, & 10 MHz	Supports of up to 40 MHz of bandwidth	1.4, 1.6, 3, 5, 10, 15, & 20 MHz	Additionally supports up to 100 MHz for downlink and 40 MHz for uplink with carrier aggregation
Spectrum	Licensed & unlicensed 2.3, 2.5, 3.5 & 5.8 GHz		Licensed IMT-2000 bands	

Framing TTI	5-ms TDD frame	0.617-ms subframes, 11 kHz subcarrier spacing, six symbols per subframe	Fixed two 0.5-ms slots = 1-ms subframes, 14 symbols per subframe	Fixed two 0.5-ms slots = 1-ms subframes, 14 symbols per subframe
Downlink pilot structure	FDM, common and dedicated pilots, supports up to two streams	FDM, common, and dedicated pilots	TDM, common, and dedicated pilots	Dedicated pilot support of up to eight streams, CSI-RS support for eight antennas
Number of codewords, CQI and ACK/NACK for downlink	One codeword, one composite CQI, and ACK/NACK	One codeword, one composite CQI, and ACK/NACK	Two codewords, per stream CQI and ACK/NACK	Same as for LTE Rel-8, differences due to carrier aggregation
Downlink sub-channelization	PUSC/band AMC	Contiguous and distributed, additional FFR zone	Localized and distributed allocation	Localized and distributed allocation
Uplink pilot structure	TDM/FDM	TDM/FDM	TDM	TDM
Uplink control channel	Data and control transmitted together	Data and control transmitted together	FDM, data and control are not transmitted together	Data and control can be transmitted together

Table 7.2. (cont.)

Feature	IEEE 802.16e (WiMAX R1)	IEEE 802.16m (WiMAX R2)	LTE Rel-8	LTE Rel-10
MCS tables	Limited and less granular	16 MCSs, larger packet and encoder block size	Highly granular	Highly granular
Uplink power control	Open-loop power control with closed-loop corrections	Open-loop power control with closed-loop corrections	Fractional OL PC with closed-loop correction, inter-cell interference mitigation using X2 interface	Same as for LTE Rel-8, modifications to support carrier aggregation
HARQ	Chase only, $N=3$	IR supported, $N=7$	Chase and IR, $N=8$ for FDD, N varies for TDD	
Total overhead	Inversely proportional to bandwidth; for 10 MHz, downlink overhead ~43%–46%	Downlink overhead <30%	Downlink overhead ~31%–33%	Downlink overhead 25%–28%

Table 7.3. *WiMAX peak data rate (29:18 TDD ratio)*

Link type	Bandwidth	WiMAX	Peak rate (Mbps)
Downlink	10 MHz	IEEE 802.16e	31.7
	10 MHz	IEEE 802.16m, 2×2	37.2
	10 MHz	IEEE 802.16m, 4×4	71.0
	10 MHz	IEEE 802.16m, 8×8	142.0
	20 MHz	IEEE 802.16m, 2×2	79.4
	20 MHz	IEEE 802.16m, 4×4	151.5
	20 MHz	IEEE 802.16m, 8×8	303.0
Uplink	10 MHz	IEEE 802.16e	5.0
	10 MHz	IEEE 802.16m, 1×2	13.1
	10 MHz	IEEE 802.16m, 2×2	24.2
	10 MHz	IEEE 802.16m, 4×4	46.4
	20 MHz	IEEE 802.16m, 1×2	27.5
	20 MHz	IEEE 802.16m, 2×2	51.8
	20 MHz	IEEE 802.16m, 4×4	99.4

Table 7.4. *LTE peak data rate (TDD configuration 1)*

Link type	Bandwidth	LTE and LTE-A	Peak rate (Mbps)
Downlink	20 MHz	2×2 SU-MIMO	82.9
	20 MHz	4×4 SU-MIMO	164.8
	20 MHz	8×8 SU-MIMO (Rel-10)	329.5
	40 MHz	2×2 SU-MIMO (Rel-10)	165.8
	40 MHz	4×4 SU-MIMO (Rel-10)	329.6
	40 MHz	8×8 SU-MIMO (Rel-10)	659.0
Uplink	20 MHz	1×2 SIMO	30.2
	20 MHz	2×2 SU-MIMO (Rel-10)	60.3
	20 MHz	4×4 SU-MIMO (Rel-10)	119.8
	40 MHz	1×2 SIMO (Rel-10)	60.4
	40 MHz	2×2 SU-MIMO (Rel-10)	120.6
	40 MHz	4×4 SU-MIMO (Rel-10)	239.6

from a marketing point of view. Note that the significant increase in uplink peak data rate on going from 802.16e (5 Mbps) to 802.16m (13.1 Mbps) is due to the availability of 64-QAM modulation in 802.16m.

Aspects of the system performance of WiMAX-and LTE-based technologies for a TDD system for two transmit antennas, two transmit/four receive antennas, two receive antennas downlink and one transmit/two receive antennas uplink are summarized in Tables 7.5 and 7.6 respectively. A 19-cell, three-sector system with full-buffer traffic using one cell reuse and three cell reuses is simulated. The notation $(1 \times 3 \times 1)$ refers to single cell reuse, with one cell having three sectors and a single frequency being reused in all the sectors. The notation $(1 \times 3 \times 3)$ refers to three-cell reuse with one cell having three sectors and three distinct carrier frequencies being used in each of the three sectors.

The following observations can be made from the comparison charts in the above tables.

- The performance of WiMAX Rel-1 is significantly inferior to that of LTE Rel-8.
- The performance of the $1 \times 3 \times 3$ reuse system using 30 MHz of bandwidth is approximately 2–2.5 times better than that of the $1 \times 3 \times 1$ reuse system using 10 MHz of bandwidth.
- The performance of 802.16m will be comparable to that of LTE Rel-8 for 2×2 downlink and 1×2 uplink and 4×2 LTE-A downlink.
- There is a significant improvement in performance with LTE-A downlink using MU/SU-MIMO (transmission mode 8/9) compared with LTE Rel-8 using 4×2 SU-MIMO (transmission mode 4) and single-layer beamforming (transmission mode 7).

The performance shown above is for a full-buffer traffic model using a specific channel model. The relative performance of LTE and WiMAX technologies is dependent upon the type of traffic, channel model, multi-antenna scheme, scheduler algorithm etc.

7.4 Migration and co-existence scenarios

WiMAX Rel-1 technology has been widely deployed around the world. Currently, 3GPP operators are doing trials and deploying Rel-8 LTE

Table 7.5. *WiMAX vs. LTE comparison for 2 × 2 downlink (DL) and 1 × 2 uplink (UL) TDD system at 10 MHz*

Parameter		WiMAX (1 × 3 × 1)	WiMAX (1 × 3 × 3)	802.16m (1 × 3 × 1)	802.16m (1 × 3 × 3)	LTE (Rel-8) (1 × 3 × 1)	LTE (Rel-8) (1 × 3 × 3)	LTE-A (Rel-10) (1 × 3 × 1)	LTE-A (Rel-10) (1 × 3 × 3)
Spectrum usage (MHz)		10 TDD 60/40 DL/UL	3 × 10 TDD 60/40 DL/UL	10 TDD 60/40 DL/UL	3 × 10 TDD 60/40 DL/UL	10 TDD configuration 1	3 × 10 TDD configuration 1	10 TDD configuration 1	3 × 10 TDD configuration 1
Antenna configuration (transmit × receive)	DL 2 × 2 UL 1 × 2	2 × 2 1 × 2	2 × 2 1 × 2	2 × 2 1 × 2	2 × 2 1 × 2	2 × 2 1 × 2	2 × 2 1 × 2	2 × 2 1 × 2	2 × 2 1 × 2
DL peak rate (Mbps) (sustained)		32	32	44	44	41.8	41.8	41.8	41.8
DL sector throughput (Mbps) DL		5.8	15.3	9.9	21.7	9.5	20.2	9.5	20.2
5% throughput (Mbps)		0.11	0.39	0.23	0.61	0.244	0.597	0.244	0.597
DL SE (bps/Hz per sector)		1.0	0.85	1.59	1.16	1.67	1.18	1.67	1.18
DL 5% SE (bps/Hz per sector)		0.018	0.021	0.037	0.032	0.0427	0.034	0.0427	0.034
UL peak rate (Mbps) (sustained)		5	5	26	26	14.6	14.6	29.2	29.2
UL sector throughput (Mbps)		1.5	3.3	2.9	4.8	3	4.8	3	4.8
UL SE (bps/Hz per sector)		0.39	0.29	0.75	0.41	0.76	0.41	0.76	0.41

SE, spectral efficiency.

Table 7.6. *WiMAX vs. LTE comparison for 4 × 2 downlink (DL) TDD system at 10 MHz*

Parameter	WiMAX (1×3×1)	WiMAX (1×3×3)	802.16m (1×3×1)	802.16m (1×3×3)	LTE (Rel-8) (1×3×1)	LTE (Rel-8) (1×3×3)	LTE- (Rel-10) (1×3×1)	LTE-A (Rel-10) (1×3×3)
Spectrum usage (MHz)	10 (TDD, 60/40 DL/UL)	3×10 (TDD, 60/40 DL/UL)	10 (TDD, 60/40 DL/UL)	3×10 (TDD, 60/40 DL/UL)	10 (TDD, configuration 1)	3×10 (TDD, configuration 1)	10 (TDD, configuration 1)	3×10 (TDD, configuration 1)
Antenna configuration (transmit/receive) DL 4×2 UL 1×4	4×2 / 1×4	4×2 / 1×4	4×2 / 2×4	4×2 / 2×4	4×2 / 1×4	4×2 / 1×4	4×2 / 2×4	4×2 / 2×4
DL peak rate (Mbps) (sustained)	32	32	44	44	41.8	41.8	41.8	41.8
DL sector throughput (Mbps)	7.4	17.6	15	29.7	8.94	18.2	13.87	21.9
DL 5% throughput (Mbps)	0.143	0.539	0.406	0.928	0.172	0.48	0.356	0.727
DL sector SE (bps/Hz per sector)	1.24	0.98	2.4	1.58	1.56	1.10	2.43	1.27
DL 5% SE (bps/Hz per sector)	0.024 (MIMO)	0.03 (MIMO)	0.065 (MU-MIMO)	0.049 (SU-MIMO)	0.03 (SU-MIMO)	0.03 (SU-MIMO)	0.062 (MU-MIMO)	0.042 (MU-MIMO)

FDD- and TDD-based technology. The large-scale commercial deployment of LTE Rel-8 technology will occur in the 2011–2012 timeframe, with deployment of LTE-A in the 2013–2015 timeframe. Operators are also exploring the migration path of their WiMAX network. Two migration scenarios are being considered – concurrent deployment of WiMAX and LTE-TDD technology and migration of WiMAX to LTE-TDD or 802.16m.

Under concurrent deployment of WiMAX and LTE-TDD, LTE-TDD will co-exist with an adjacent WiMAX carrier. WiMAX Rel-1 has been deployed in a $1 \times 3 \times 3$, $1 \times 4 \times 2$ or $1 \times 4 \times 4$ reuse pattern. Deployment of WiMAX in a $1 \times 3 \times 1$ pattern using fractional frequency reuse and inter-cell interference-coordination schemes is being considered. For this scenario to be feasible, the downlink/uplink split for these two technologies should be similar in order to prevent adjacent-carrier interference between LTE-TDD and WiMAX. The most common downlink/uplink ratio for WiMAX Rel-1 using a bandwidth of 10 MHz is the 60:40 configuration, meaning that 29 OFDMA symbols are allocated to downlink transmission and 18 OFDMA symbols are allocated to uplink transmission out of a total of 47 symbols. The 29 downlink symbols consist of 1 preamble symbol and 14 PUSC data slots, which occupy 2 OFDMA symbols each. In the uplink, the 18 symbols typically consist of 3 control symbols carrying the random-access channel and feedback channels and 5 PUSC data slots. Unlike in the downlink, the uplink PUSC data slots occupy three OFDMA symbols. The TD-LTE frame structure and the available downlink/uplink are given in Section 3.3. In order to time-align an LTE-TDD with WiMAX, the LTE-TDD system should use a switching period of 5 ms (LTE-TDD configurations 0, 1, 2, and 6) as well as a similar downlink and uplink transmission period. Of the available downlink/uplink LTE-TDD configurations, configuration 1 has similar transmission periods to WiMAX with a downlink/uplink ratio of 60:40. The LTE-TDD special subframe format 4 is used to provide an approximate match to the WiMAX 60:40 frame with only a 2% overlap between BS and MS transmission periods. Figure 7.1 illustrates the compatible LTE-TDD and WiMAX frame structures. Note that the LTE-TDD radio frame starts 1 ms later than the WiMAX frame. The simplest solution to eliminate the remaining 2% overlap between the downlink transmission period of WiMAX and the uplink transmission period of LTE-TDD is to

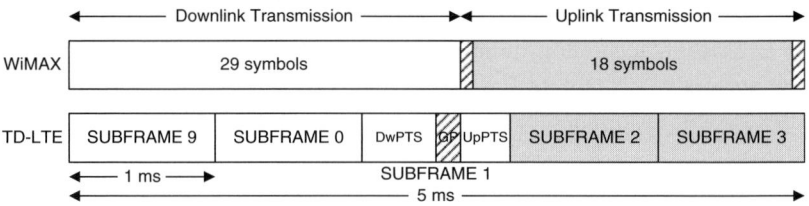

Figure 7.1. LTE-TDD and WiMAX frame structures for concurrent operation.

drop two downlink WiMAX symbols. This results in a reduction in maximum throughput of 7% (=2/29) to the WiMAX downlink with no loss to the LTE-TDD system. For deployment in this mode, a UE should be capable of supporting both LTE-TDD and WiMAX technology including single- or dual-mode devices.

When migrating from WiMAX to LTE-TDD, the network supports only LTE-TDD-capable devices and the radio access network is fully converted into an LTE-TDD network. Only LTE-TDD-capable UE or dual-mode devices can be supported. The migration from WiMAX Rel-1 to 802.16m is more natural, since 802.16m is fully backward compatible with respect to WiMAX Rel-1.

Additional reading

[1] Chang, M. J., Abichar, Z., Chau-Yun, H., "WiMAX or LTE: who will lead the broadband mobile internet?," *IT Professional*, vol. 12, no. 3, pp. 26–32, May–June 2010.

[2] Ball, C., Hindelang, T., Kambourov, I., Eder, S., "Spectral efficiency assessment and radio performance comparison between LTE and WiMAX," *IEEE 19th International Symposium on Personal, Indoor and Mobile Radio Communications*, September 2008.

[3] Krapichler, C., "LTE, HSPA and mobile WiMAX: a comparison of technical performance," *Institution of Engineering and Technology Hot Topics Forum: LTE vs WiMAX and Next Generation Internet*, September 2007.

[4] Srikanth, S., Pandian, P. A. M., "Orthogonal frequency division multiple access in WiMAX and LTE – a comparison," *National Conference on Communications (NCC)*, January 2010.

[5] Zhijie, W., Yafeng, W., Fei, W., "Comparison of VoIP capacity between 3G-LTE and IEEE 802.16m," *IEEE 20th International Symposium on Personal, Indoor and Mobile Radio Communications*, pp. 2192–2196, September 2009.

[6] Lowe, S., "LTE vs WiMAX," *Institution of Engineering and Technology Hot Topics Forum: LTE vs WiMAX and Next Generation Internet*, September 2007.

[7] Pulley, D., "Infrastructure implementation challenges for LTE and WiMAX air interfaces," *Institution of Engineering and Technology Hot Topics Forum: LTE vs WiMAX and Next Generation Internet*, September 2007.

[8] Report ITU-R M. 2135, "Guidelines for evaluation of radio interface technologies for IMT-Advanced," 2008.

[9] Report ITU-R M. 2134, "Requirements related to technical performance for IMT-Advanced radio interface(s)," 2008.

[10] 3G Americas, *3GPP Mobile Broadband Innovation Path to 4G: Release 9/10 and Beyond*, February 2010.

[11] Rysavy Research, *HSPA to LTE-Advanced: 3GPP Broadband Evolution to IMT-Advanced*, September 2009.

[12] WiMAX Forum, *Mobile WiMAX–Part I/II: A Technical Overview and Performance Evaluation*, February 2006.

Appendix

A.1 System analysis and performance metrics

In this book, system-level performance results based on comprehensive system simulations of cellular networks are provided. An example of the cellular layout used for system simulation is shown in Figure A.1. This is a typical 19-site, 57-cell system using a hexagonal grid. In this case, a cell is viewed as a sector of the physical site. However, in LTE each cell is treated as an independent eNB. The spacing between each site and the next is dependent on the deployment scenario. For example, in urban micro-cell deployment, the inter-site distance is 200 m. Users are dropped randomly into the simulation space. For instance, in urban micro-cell deployment, 570 users are randomly dropped. After the users have been dropped, long-term radio characteristics such as pathloss and shadowing are calculated. Users are then assigned to the cell using the minimum pathloss as the cell-selection criterion. For the urban micro-cell example, on average, approximately 10 users will be associated with each cell.

Next, system analysis is performed as the system operates according to the standards. The basic analysis steps include scheduling, transmission-mode selection and resource allocation, mapping to the physical layer, transmission, link-error prediction, and HARQ. The simulations are run for a fixed number of user drops and time. Afterwards, the following physical-layer performance metrics are collected and reported.

- Cell/sector throughput, defined as the total number of over-the-air (i.e. over-the-physical-layer) information bits that were successfully delivered to or from the cell within the simulation time. This number is averaged over all the cells in the system.
- User throughput, defined as the total number of over-the-air information bits that were successfully delivered within the transmission time for a user. This statistic is determined for all the users in the system.

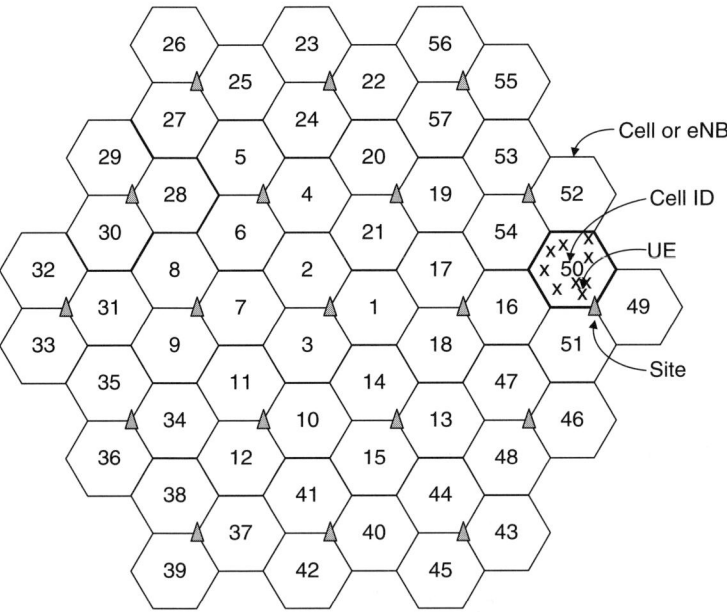

Figure A.1. Cellular layout for system simulation.

- The 95th-percentile user throughput, defined as the user throughput at the 95th percentile of the sorted user throughput distribution. For example, in a 57-cell system with 570 users and 10 simulation drops, there are 5700 user throughput data points. These values are first sorted into ascending order and the 95th-percentile throughput is given by the throughput of the 5415th user within the sorted user distribution. This metric is also sometimes referred to as cell-center user throughput.
- Cell-edge user throughput (5th-percentile throughput), defined as the user throughput at the 5th percentile of the sorted user throughput distribution. Using the above example, the cell-edge throughput is given by the throughput of the 285th user within the sorted user distribution.

The unit for the above performance metrics is bits per second (bps). To provide an easy basis for comparison between systems with different bandwidths, these results are usually normalized by the system bandwidth to give spectral efficiencies. For instance, if the cell throughput is

11 Mbps and the system bandwidth is 10 MHz, the cell spectral efficiency is 1.1 bits per second per Hz (bps/Hz).

Scheduling refers to the selection of users that will be assigned time–frequency resources in the next subframe. Several scheduling methods are available. Typical scheduling methods include round-robin, maximum SINR, and proportional fairness. In round-robin scheduling, users take turns being scheduled. This type of scheduling has a high degree of fairness insofar as all users are equally scheduled regardless of their channel conditions. As a result, the cell-edge user throughput is high but system performance with round-robin scheduling is poor. On the other hand, in maximum-SINR scheduling, users are ranked according to their current channel conditions, and only users with good channel conditions are scheduled. In this case, the level of fairness is poor since users in bad channel conditions might not be assigned any resources at all. Maximum-SINR scheduling results in very high system throughput but very poor cell-edge throughput. Proportional fair scheduling attempts to balance fairness and throughput by limiting how often good users are scheduled on the basis of fairness to all users. In this case, each user is assigned a metric called the proportional fair (PF) metric. In every subframe, users are sorted according to this metric and the user with the highest value is the first to be given resources. The components of the PF metric include average achieved throughput, possible throughput if scheduled, fairness factors, and a weight factor. Average throughput refers to the actual data rate this user is experiencing, while possible throughput refers to the possible data rate that can be achieved by this user in this subframe. In addition, two fairness factors can be used to adjust the priority to be assigned to each of these components. Finally, an additional weight factor can be used to modify the priority depending on other criteria independent of throughput (e.g. packet delay or service type). For example, VoIP traffic may be given higher priority than HTTP through the use of this weight factor.

Resource allocation determines how many resource blocks can be assigned to this user, what kind of MIMO scheme to use, the transport-block size, the power level, and the modulation and coding rate. Together with scheduling, resource allocation is paramount to achieving good performance. In general, resource allocation is done using a greedy

Table A.1. *System simulation parameters*

Parameter	Assumption
Cellular layout	Hexagonal grid, 19 sites, 3 cells per site
Antenna pattern	Directional antenna with beam-width 70° and maximum attenuation 20 dB
Channel model	IMT channel model
HARQ	IR with stop-and-wait HARQ protocol
User-selection metric	Proportional fair
Link-error-prediction method	Exponential effective SIR mapping (EESM)
Other cell interference	Explicitly modeled
Control channels	Explicitly modeled
CQI mode	Periodic, modes 1–1 and 2–1
CQI delay	5 ms
Channel estimation	Non-ideal

approach. For example, in the uplink, once a user has been selected for scheduling, he is allowed to take as many resource blocks as possible subject to a power constraint. This is also true in the downlink when frequency-non-selective scheduling is used.

Table A.1 list pertinent system simulation parameters used to generate the results in this book. They include items of information related to the channel model, antenna pattern, link-error-prediction method, overhead assumptions, and implementation details. For example, Table A.1 shows the CQI mode and associated delay assumed when evaluating downlink system performance.

A.2 Abbreviations

ABS	Almost blank subframes
ACK	Acknowledgment

AMC	Adaptive modulation and coding
ARQ	Automatic repeat request
AS	Access stratum
BCCH	Broadcast Control Channel
BCH	Broadcast Channel
BSR	Buffer status report
CA	Carrier aggregation
CAZAC	Constant-amplitude zero auto-correlation
CCE	Control-channel element
CDD	Cyclic-delay diversity
CFI	Control-format indicator
CIF	Carrier-indicator field
CP	Cyclic prefix
C-plane	Control plane
CQI	Channel quality indicator
CRC	Cyclic redundancy check
C-RNTI	Cell RNTI
CSG	Closed subscriber group
CSI-RS	Channel state information RS
DAI	Downlink assignment index
DAS	Distributed antenna system
DCCH	Dedicated Control Channel
DCI	Downlink control information
DFT	Discrete Fourier transform
DFT-S-OFDM	DFT spread OFDM
DL	Downlink
DL-SCH	Downlink shared channel
DRX	Discontinuous reception
DTCH	Dedicated Traffic Channel
DTX	Discontinuous transmission
DwPTS	Downlink pilot time slot
eNB	E-UTRAN node B
EPC	Evolved packet core
EPRE	Energy per resource element
E-UTRA	Evolved UTRA

E-UTRAN	Evolved UMTS terrestrial radio access network
FDD	Frequency-division duplexing
FFT	Fast Fourier transform
FDM	Frequency-division multiplexing
HARQ	Hybrid ARQ
HI	HARQ indicator
HO	Handover
HSDPA	High-speed downlink packet access
ICIC	Inter-cell interference coordination
IoT	Interference over thermal
IP	Internet Protocol
IR	Incremental redundancy
LTE	Long Term Evolution
LTE-A	LTE-Advanced
MAC	Medium-access control
MBMS	Multimedia broadcast multicast service
MCH	Multicast channel
MCS	Modulation and coding scheme
MIB	Master information block
MIMO	Multiple-input multiple-output
MME	Mobility management entity
MU-MIMO	Multi-user MIMO
NACK	Negative acknowledgment
NAS	Non-access stratum
OFDM	Orthogonal frequency-division multiplexing
OFDMA	Orthogonal frequency-division multiple access
PA	Power amplifier
PAPR	Peak-to-average power ratio
PBCH	Physical Broadcast Channel
PCCH	Paging Control Channel
PCFICH	Physical Control Format Indicator Channel
PCH	Paging Channel
PCI	Physical cell identifier
PDCCH	Physical Downlink Control Channel
PDCP	Packet Data Convergence Protocol

PDSCH	Physical Downlink Shared Channel
PDU	Protocol data unit
P-GW	PDN gateway
PHICH	Physical Hybrid ARQ Indicator Channel
PHR	Power headroom report
PHY	Physical layer
PMCH	Physical Multicast Channel
PMI	Precoding-matrix indicator
PRACH	Physical Random Access Channel
PRB	Physical resource block
P-RNTI	Paging RNTI
PSS	Primary synchronization signal
PUCCH	Physical Uplink Control Channel
PUSCH	Physical Uplink Shared Channel
QAM	Quadrature amplitude modulation
QCI	QoS class identifier
QoS	Quality of service
RACH	Random Access Channel
RA-RNTI	Random-access RNTI
RAT	Radio access technology
RAT	Resource-allocation type
RB	Resource block
RBG	Resource-block group
RE	Resource element
RF	Radio frequency
RI	Rank indication
RLC	Radio link control
RN	Relay node
RNC	Radio network controller
RNTI	Radio network temporary identifier
ROHC	Robust header compression
RPF	Repetition factor
RRC	Radio resource control
RRH	Remote radio head
RRM	Radio resource management

RS	Reference signal
RSCP	Received signal code power
RSRP	Reference-signal received power
RSRQ	Reference-signal received quality
RSSI	Received-signal strength indicator
S1-MME	S1 for the control plane
S1-U	S1 for the user plane
SAE	System architecture evolution
SC-FDMA	Single-carrier–frequency-division multiple access
SCH	Synchronization channel
SDMA	Spatial-division multiple access
SDU	Service data unit
SFBC	Space frequency block code
SFN	Single-frequency network
SFN	System frame number
S-GW	Serving gateway
SI	System information
SIB	System-information block
SIMO	Single-input multiple-output
SINR	Signal-to-interference-plus-noise ratio
SIR	Signal-to-interference ratio
SI-RNTI	System-information RNTI
SPS	Semi-persistent-scheduling
SPS C-RNTI	Semi-persistent scheduling C-RNTI
SR	Scheduling request
SRS	Sounding reference symbols
SSS	Secondary synchronization signal
SU-MIMO	Single-user MIMO
TA	Time alignment
TB	Transport block
TBS	Transport-block size
TCP	Transmission control protocol
TDD	Time-division duplexing
TPC	Transmit power control
TPMI	Transmitted precoding-matrix indicator

TTI	Transmission time interval
UCI	Uplink control information
UE	User equipment
UL	Uplink
UL-SCH	Uplink shared channel
UMTS	Universal Mobile Telecommunication System
U-plane	User plane
UpPTS	Uplink pilot time slot
UTRA	Universal Terrestrial Radio Access
UTRAN	Universal Terrestrial Radio Access Network
VRB	Virtual resource block
WCDMA	Wideband code-division multiple access

Index